Lecture Notes in Mathematics

Edited by A. Dold, F. Takens and B. Teissier

Editorial Policy
for the publication of monographs

1. Lecture Notes aim to report new developments in all areas of mathematics – quickly, informally and at a high level. Monograph manuscripts should be reasonably self-contained and rounded off. Thus they may, and often will, present not only results of the author but also related work by other people. They may be based on specialized lecture courses. Furthermore, the manuscripts should provide sufficient motivation, examples and applications. This clearly distinguishes Lecture Notes from journal articles or technical reports which normally are very concise. Articles intended for a journal but too long to be accepted by most journals, usually do not have this "lecture notes" character. For similar reasons it is unusual for doctoral theses to be accepted for the Lecture Notes series.

2. Manuscripts should be submitted (preferably in duplicate) either to one of the series editors or to Springer-Verlag, Heidelberg. In general, manuscripts will be sent out to 2 external referees for evaluation. If a decision cannot yet be reached on the basis of the first 2 reports, further referees may be contacted: the author will be informed of this. A final decision to publish can be made only on the basis of the complete manuscript, however a refereeing process leading to a preliminary decision can be based on a pre-final or incomplete manuscript. The strict minimum amount of material that will be considered should include a detailed outline describing the planned contents of each chapter, a bibliography and several sample chapters.
Authors should be aware that incomplete or insufficiently close to final manuscripts almost always result in longer refereeing times and nevertheless unclear referees' recommendations, making further refereeing of a final draft necessary.
Authors should also be aware that parallel submission of their manuscript to another publisher while under consideration for LNM will in general lead to immediate rejection.

3. Manuscripts should in general be submitted in English.
Final manuscripts should contain at least 100 pages of mathematical text and should include
- a table of contents;
- an informative introduction, with adequate motivation and perhaps some historical remarks: it should be accessible to a reader not intimately familiar with the topic treated;
- a subject index: as a rule this is genuinely helpful for the reader.

Lecture Notes in Mathematics 1719

Springer
Berlin
Heidelberg
New York
Barcelona
Hong Kong
London
Milan
Paris
Singapore
Tokyo

Kenneth R. Meyer

Periodic Solutions of the N-Body Problem

Springer

Author

Kenneth R. Meyer
Department of Mathematics
University of Cincinnati
Cincinnati, Ohio 45221-0025
USA
E-mail: ken.meyer@uc.edu

Cataloging-in-Publication Data applied for

Die Deutsche Bibliothek - CIP-Einheitsaufnahme

Meyer, Kenneth R.:
Periodic solutions of the N-body problem / Kenneth R. Meyer. - Berlin ; Heidelberg ; New York ; Barcelona ; Hong Kong ; London ; Milan ; Paris ; Singapore ; Tokyo : Springer, 1999
(Lecture notes in mathematics ; 1719)
ISBN 3-540-66630-3

Mathematics Subject Classification (1991): 58F05, 58F22, 70F10, 70F15

ISSN 0075-8434
ISBN 3-540-66630-3 Springer-Verlag Berlin Heidelberg New York

Typesetting: Camera-ready TEX output by the author
SPIN: 10700296 41/3143-543210 - Printed on acid-free paper

Preface

These notes grew out of a series of lectures that I gave at the Universidade Federal de Pernambuco, Recife, Brazil. Since this was a limited number of lectures in the extensive area of periodic solutions of the N-body problem, I was forced to define a small subset of the literature in order to give a reasonably complete introduction. Filling in the most of the details resulted in these lecture notes.

From a generic point of view the N-body problem is highly degenerate. It is invariant under the symmetry group of Euclidean motions and admits linear momentum, angular momentum and energy as integrals. This implies that an attempt to apply the implicit function directly yields a Jacobian with nullity 8 for the planar problem and nullity 12 for the spatial problem. (The multiplier $+1$ has multiplicity 8 in the planar problem and 12 in the spatial problem.) Therefore, the integrals and symmetries must be confronted head on, which leads to the definition of the reduced space where all the known integrals and symmetries have been eliminated. It is on the reduced space that one can hope for a nonsingular Jacobian without imposing extra symmetries.

The first six chapters develops the theory of Hamiltonian systems, symplectic transformations and coordinates, periodic solutions and their multipliers, symplectic scaling, the reduced space etc. The remaining six chapters contain theorems which establish the existence of periodic solutions of the N-body problem on the reduced space.

The N-body problem is the classical prototype of a Hamiltonian system with a large symmetry group and many first integrals. These lecture notes are an introduction to the theory of periodic solutions of such Hamiltonian systems.

I would like to thank Hildeberto Cabral for his kind hospitality during my visit to the Universidade Federal de Pernambuco, Recife, Brazil. The final version of this monograph was completed while I was the Fundació Banco Bilbao Vizcaya Scholar at the Centre de Recerca Matemàtica, Institut d'Estudis Catalans. Jaume Llibre and his colleagues at the Universitat Autònoma de Barcelona were most cordial and helpful.

Many people read various parts of the manuscript in its various stages and sent me comments and corrections. I would like to thank Martha Alvarez,

Hildeberto Cabral, Anne Feldman, Karl Meyer, and Gareth Roberts for their help. A special thanks goes to N. V. Fitton for her revisions of my revisions.

I am sure there are some errors in these notes and I hope they are small. Please notify me of all errors large or small at Department of Mathematics, University of Cincinnati, Cincinnati, Ohio 45221-0025, USA or ken.meyer@uc.edu.

My research has been supported by grants from the National Science Foundation and the Charles Phelps Taft Foundation.

University of Cincinnati, August 1999 *Kenneth R. Meyer*

Contents

1. Introduction

1.1 History

The N-body problem is a system of ordinary differential equations that describes the motion of N point masses or particles moving under Newton's laws of motion, where the only forces acting are the mutual gravitational attractions. The problem is solved for $N = 2$ because it can be reduced to the Kepler problem which is a system of ordinary differential equations that describes the motion of a particle moving under the gravitational attraction of a second particle fixed at the origin. The solutions of the Kepler problem are conic sections — circles, ellipses, parabolas, and hyperbolas.

Newton's formulation of his laws of motion and his law of gravity was one of the greatest scientific accomplishments of all times. With these simple principles he was able to completely solve the two-body problem deriving Kepler's laws describing the motion of the planet Mars. To the first approximation the orbit of Mars is a solution of the two-body problem where only the gravitational forces of the sun and Mars are taken into account and this problem can be reduced to the Kepler problem. Using a perturbation analysis he was able estimate some of the higher order effects and so explain some of the anomalies in Mar's orbit.

Newton next turned to the problem of describing the orbit of the moon. This is a harder problem since the first approximation should be a three-body problem — the earth, moon, and sun. The problem he encountered caused him to remark to the astronomer John Machin that "*... his head never ached but with his studies on the moon.*"[1]

Today the orbit of the moon is obtained either by numerical integration or by asymptotic series expansion — see [29].

It is now widely believed that the N-body problem for $N \geq 3$ cannot be solved in the same sense as the two-body problem. In fact there is very good evidence that the general N-body problem is not solvable. However, since Newton's time there have been thousands of papers written on the N-body problem. These papers contain special solutions, asymptotic estimates, information about collision, the existence and non-existence of integrals, series solutions, non-collision singularities, etc.

---
[1] [37]

The existence, stability and bifurcation of periodic solutions of the N-body problem has been the subject of many of these papers particularly since the works of Poincaré. Poincaré wrote extensively on periodic solutions and in particular a large portion of his *Les méthodes nouvelles de la mécanique céleste* [66] is devoted to this topic. He said about periodic solutions of the three-body problem

> En effet, il y a une probabilité nulle pour que les conditions initiales du mouvement soient précisément celles qui correspondent à une solution périodique. Mais il peut arriver qu'elles en diffèrent très peu, et cela a lieu justement dans les cas où les méthodes anciennes ne sont plus applicables. On peut alors avec avantage prendre la solution périodique comme première approximation, comme *orbite intermédiaire*, pour employer le langage de M. Gyldén.
>
> Il y a même plus: voici un fait que je n'ai pu démontrer rigoureusement, mais qui me paraît pourtant très vraisemblable.
>
> Étant données des équations de la forme définie dans le n° 13 et une solution particulière quelconque de ces équations, on peut toujours trouver une solution périodique (dont la période peut, il est vrai, être très longue), telle que la différence entre les deux solutions soit aussi petite qu'on le veut, pendant un temps aussi long qu'on le veut. D'ailleurs, ce qui nous rend ces solutions périodiques si précieuses, c'est qu'elles sont, pour ainsi dire, la seule brèche par où nous puissions essayer de pénétrer dans une place jusqu'ici réputée inabordable. [2]

This conjecture was often quoted by Birkhoff as a justification for his work on fixed point theorems and related topics — see for example [12, 13]. Poincaré conjecture that periodic orbits are dense has only been established for C^1-generic Hamiltonian systems on a compact manifold by Pugh and Robinson [68] and in a certain sense for the restricted three-body problem by Gómez and Llibre [28].

There is an extensive literature on the existence and nature of periodic solutions of the N-body problem, especially the restricted three-body problem. Many different methods have been used to establish the existence of periodic solutions in the N-body problem and Hamiltonian systems in general, for example: averaging — see Moser [55], the Lagrangian manifold intersection theory — see Weinstein [90], normal forms — see Schmidt [74], numeric — see Ángel and Simó [2], majorants — see Liapunov [43] and Siegel [80], special fixed-point theorems — see Birkhoff [13], symbolic dynamics — see Saari and Xia [71], variational methods — see Robinovich [69], and many others. This is but a small sample of a vast subject. This monograph is concerned with one small subset of the literature where I have made some contributions.

[2] [66, p. 81]

I will establish the existence and discuss linear stability of periodic solutions of the full N-body problem without exploiting a discrete symmetry. Almost all the early literature on the existence of periodic solutions uses a discrete symmetry and so only applies in certain symmetric configurations and in general does not give any stability information. Also most of this literature establishes the existence in the three-body or restricted three-body problem only. I will extend Poincaré's continuation method to new applications by using symplectic scaling and the symplectic reduction theorem.

Poincaré's continuation method is a simple perturbation method. It requires a small parameter ε, which may be a physical quantity, such as one of the masses, or a scale parameter, which measures the distance between two of the bodies. A solution is periodic if it returns to its initial position after a time T, the *period*. This results in a finite set of equations that must be solved. Poincaré's continuation method uses the finite-dimensional implicit function theorem to solve these equations. When $\varepsilon = 0$ one finds a solution, computes the necessary Jacobian to be nonsingular, applies the implicit function theorem, and concludes that the solution continues to exist when $\varepsilon \neq 0$ but small. This method is introduced in Chapter 6 and used in Chapters 7–12.

Symplectic scaling is the method of introducing the small parameter ε into the problem while preserving the Hamiltonian nature of the problem. The art is to introduce the parameter in such a way that when $\varepsilon = 0$ the problem has a periodic solution with the requisite Jacobian nonsingular, which then yields an interesting theorem. It seems that the scaling yields interesting results only when it preserves the symplectic structure. Symplectic scaling is introduced in Chapter 3 and used in Chapters 7–12.

The initial formulation of the N-body problem is a system of equations in $\mathbb{R}^{4n} \backslash \Delta$ for the planar problem or $\mathbb{R}^{6n} \backslash \Delta$ for the spatial problem where Δ is the collision set. But the correct place to study the problem in on the reduced space a $(4N-6)$-dimensional symplectic manifold for the planar problem or $(6N-10)$-dimensional symplectic manifold for the spatial problem in general. It is only on this reduced space that one can hope to study the existence and stability of periodic solutions.

It has been known since the time of Newton that the N-body problem is invariant under a Euclidean motion (a translation followed by a rotation) and admits the integrals of linear and angular momentum. They are a curse and a blessing: a curse because they make the Jacobian of Poincaré's continuation method highly singular, and a blessing since they can be used to reduce the dimension of the problem. When the symmetries and integrals are used correctly, the problem can be reduced to a much lower-dimensional problem which is again Hamiltonian. This is called the Meyer-Marsden-Weinstein reduction; it is the main result of my paper [50] and the paper of Marsden and Weinstein [44]. In the problems discussed in this monograph, after symplectic scaling and reduction the requisite Jacobian is nonsingular. Again this is the art, since many seemingly similar problems are still degenerate when all the symmetries and integrals are accounted for. Symplectic reduction is discussed in detail in Chapter 5 and used in Chapters 7–12.

Many of the chapters end with a list of problems. Some are routine and some are not. I recommend at least looking at them, since they often contain generalizations and related results. Sometimes the problem ends with a reference, in which case the reader should realize that at some point in time the solution of this problem was considered a publishable result.

1.2 Global vs. Local Notation

There is an old saying in celestial mechanics that no set of variables is good enough. The subject is replete with different sets of variables, many of which bear the names of some of the greatest mathematicians of all times. There are not enough alphabets to give each variable a separate symbol, so I will use a concept from programming languages — global and local. Some symbols will stand for the same quantity throughout the notes, *global variables*, and some will stand for different quantities in different sections, *local variables*. In either case I will say what a variable stands for in each context.

Throughout these notes fixed and rotating reference frames will be used, rotating more often than fixed. Therefore, quantities referring to a fixed reference frame will in general be in a boldface font, whereas the same quantity in a rotating frame will be in a regular font. I use a Hamiltonian formalism throughout the notes and write different Hamiltonians in many different variables. I will always use the generic symbols H, $\mathbf{H}$ for the Hamiltonian, but say something like, "The Hamiltonian of the $\cdots$ problem in $\cdots$ variables is $\cdots$."

Here is a list of global variables.

- $\mathbf{q}, \mathbf{p}$ are the position and momentum vectors in fixed rectangular coordinates.
- q, p are the position and momentum vectors in rotating rectangular coordinates.
- $\mathbf{x}, \mathbf{y}$ are the position and momentum vectors in fixed Jacobi coordinates.
- x, y are the position and momentum vectors in rotating Jacobi coordinates.
- $\mathbf{H}$ is the current Hamiltonian in fixed coordinates.
- H is the current Hamiltonian in rotating coordinates.
- $\mathbf{O}$ is the current angular momentum vector in fixed coordinates.
- O is the current angular momentum vector in rotating coordinates.
- $\mathbf{U}$ is the current (self-) potential in fixed coordinates.
- U is the current (self-) potential in rotating coordinates.
- ε is the current perturbation parameter.
- m_i is the mass of the ith particle.

Many of the above global variables will be subscripted.

The variables $u, v, \xi, \zeta, \alpha, \beta$, etc. are local variables whose meanings will be given in context.

$\mathbb{R}$ will denote the field of real numbers, $\mathbb{C}$ the complex field, and $\mathbb{F}$ either $\mathbb{R}$ or $\mathbb{C}$. $\mathbb{F}^n$ will denote the space of all n-dimensional column vectors, and, unless otherwise said, all vectors will be column vectors. However, vectors will be written as row vectors within the body of the text for typographical reasons. $\mathcal{L}(\mathbb{F}^n, \mathbb{F}^m)$ will denote the set of all linear transformations from $\mathbb{F}^n$ to $\mathbb{F}^m$ and will sometimes be identified with the set of all $m \times n$ matrices.

If A is a matrix, then A^T will denote its transpose, A^{-1} its inverse, and A^{-T} the inverse transpose, if these matrices exist. A matrix A is block diagonal if it is of the form

$$A = \begin{pmatrix} A_{11} & O_{12} & O_{13} & \cdots & O_{1k} \\ O_{21} & A_{22} & O_{23} & \cdots & O_{2k} \\ O_{31} & O_{32} & A_{33} & \cdots & O_{3k} \\ & \cdots & & \cdots & \\ O_{k1} & O_{k2} & O_{k3} & \cdots & A_{kk} \end{pmatrix}$$

where the A_{ii} are square matrices and the O_{ij} are the rectangular zero matrices. We will write $A = \operatorname{diag}(A_{11}, A_{22}, ..., A_{kk})$.

Functions will be real and smooth unless otherwise said, where smooth means C^∞ or real analytic. If $f(u)$ is a smooth function from an open set $\mathcal{O}$ in $\mathbb{R}^n$ into $\mathbb{R}^m$, then $\partial f/\partial u$ will denote the $m \times n$ Jacobian matrix

$$\frac{\partial f}{\partial u} = \begin{pmatrix} \frac{\partial f_1}{\partial u_1} & \cdots & \frac{\partial f_1}{\partial u_n} \\ & \cdots & \\ & \cdots & \\ \frac{\partial f_m}{\partial u_1} & \cdots & \frac{\partial f_m}{\partial u_n} \end{pmatrix}$$

If $f : \mathbb{R}^n \to \mathbb{R}^1$, then $\partial f/\partial u$ is a row vector; let ∇f or $\nabla_u f$ or f_u denote the column vector $(\partial f/\partial u)^T$. (Even when discussing functions on manifolds, no Riemannian metric is assumed, so ∇f is not the gradient vector with reference to some Riemannian matrix.) When the derivative of f is thought of as a map from $\mathcal{O}$ into $\mathcal{L}(\mathbb{R}^n, \mathbb{R}^m)$, the space of linear operators from $\mathbb{R}^n$ to $\mathbb{R}^m$, the derivative will be denoted by Df. The variable t will denote a real scalar variable called time, and we use Newtonian dots for the first and second derivatives, i.e., $\dot{} = d/dt$, and $\ddot{} = d^2/dt^2$.

1.3 Summary of Chapters

Chapters 2–6 give general background material on Hamiltonian systems, the N-body problem, symplectic manifolds, periodic solutions, etc. People with a good knowledge of these topics can skim through these chapters quickly. These chapters assume a knowledge of the basic theory of ordinary differential equations, i.e., existence, uniqueness, linear theory, etc. Also, from time to time proofs are given by reference.

Each of Chapters 7–12 contains a theorem establishing the existence of a class of periodic solutions of the N-body problem. These later chapters depend on the earlier chapters, but they are independent of each other; and while each has its peculiarities, they are very similar in form. Therefore, the reader should choose which chapters to read. Chapters 7 and 8 are the simplest, and are therefore recommenced for the first reading.

Here is a summary of the chapters.

Chapter 2 This chapter introduces the N-body problem as a Hamiltonian system of equations. The classic integrals of energy, linear momentum, and angular momentum are derived. Then some of the special cases are given, namely, the Kepler problem (central force problem), the restricted three-body problem, Hill's lunar equation, and the elliptic restricted three-body problem. All of the systems are given as examples of Hamiltonian systems.

Chapter 3 Now that we have seen some examples, it is time to give some basic theory of Hamiltonian systems — at least as it pertains to celestial mechanics and periodic solutions. The changes of variables that preserve the Hamiltonian character of the problem are called symplectic. We give the basic definition of symplectic changes of variables along with the main examples of symplectic variables to be used later.

Chapter 4 A central configuration is a configuration of N particles giving rise to special solutions of the N-body problem. In one such solution coming from a central configuration, all the particles uniformly rotate about their center of mass while maintaining their relative positions. Such a solution is called a relative equilibrium. For example, there is a periodic solution of the three-body problem in which three particles remain at the vertices of an equilateral triangles while uniformly rotating. This chapter introduces central configurations and gives a special coordinate system for central configurations.

Chapter 5 The N-body problem has many symmetries due to the facts that the particles are assumed to be point masses and Newton's law of gravity assumes that space is homogeneous and isotropic. Symmetries often introduce degenerates, which can cause problems with the analysis unless the symmetries are exploited correctly. This chapter is devoted to understanding the main symmetry of the N-body problem, that is, its invariance under the group of Euclidean motions. The basic result discussed is the symplectic reduction theorem, which states that when all the classic integrals are held fixed and all the remaining symmetries are removed, the resulting system is again Hamiltonian. It is on this reduced space that the perturbation analysis yields the periodic solutions.

Chapter 6 This chapter develops Poincaré's continuation method for systems with various degrees of degeneracies. The main example is the N-body problem.

Chapter 7 Here we prove the existence of Poincaré's "periodic orbits of the first kind" by the methods developed in the previous chapters. We establish periodic solutions in which two particles assumed to have small mass, called the satellites, move on an approximately circular orbit about a particle of large mass, called the primary. These are the simplest examples of periodic solutions on the reduced space.

Chapter 8 We show that under mild nonresonance assumptions, that a nondegenerate periodic solution of the restricted problem can be continued into the full $(N+1)$-body problem. This result follows easily from the Hamiltonian of the $(N+1)$-body problem with one small mass, after it has been correctly scaled. This is the easiest example of symplectic scaling, which shows that the restricted problem is indeed the first approximation of the full problem with one small mass.

Chapter 9 In this chapter we show that there are periodic solutions of the $(N+1)$-body problem on the reduced space in which $N-1$ particles and the center of mass of the other two particles move approximately on a relative equilibrium solution, while the other two particles move approximately on a small circular orbit of the two-body problem about their center of mass.

Chapter 10 The main result of this chapter is the existence of a family of periodic solutions of the planar $(N+1)$-body problem on the reduced space in which one particle, called the comet, is at a great distance from the other N particles, called the primaries. The comet moves approximately on a circular orbit of the Kepler problem about the center of mass of the primary system, and the primaries move approximately on a relative equilibrium solution. This is the most degenerate of all the problems discussed in these notes.

Chapter 11 In this chapter we use the method of symplectic scaling of the Hamiltonian in order to give a precise derivation of the main problem of lunar theory. Under one set of assumptions, we derive the main problem used by Delaunay, and under another, the main problem as given by Hill. The derivations are precise asymptotic statements about the limiting behavior of the three-body problem and so can be used to give precise estimates of the deviation of the solutions of the first approximation and the full solutions. Using this scaling, we prove that any nondegenerate periodic solution of Hill's lunar equations whose period is not a multiple of 2π can be continued into the full three-body problem on the reduced space.

Chapter 12 This chapter deals with the planar N-body problem of classical celestial mechanics and its relation to the elliptic restricted problem. This system, unlike the previous system, is periodic. We give a different derivation of the elliptic restricted problem which gives a restricted problem for each type of solution of the Kepler problem. Again we show that any nondegenerate periodic solutions of the elliptic restricted problem whose period is not a multiple of 2π can be continued into the full three-body problem on the reduced space.

1.4 Further Reading

This book assumes some knowledge of basic differential equations as found, for example, in the introductory texts by Sánchez [72] or Arrowsmith and Place [9]. These are readable, short introductions to the geometric theory of differential equations and they should give sufficient background. More advanced texts are [31, 32]. References to special advanced topics will be given as needed.

Pollard [67] gives a clean and complete description of the solution to the two-body problem, an introduction to Hamiltonian equations, and a brief treatment of the restricted problem. This short book is an ideal starting point for the study of Hamiltonian systems and celestial mechanics. A more elementary and classical introduction is found in Danby [22] or Moulton [58].

At a higher level of difficulty are: Meyer and Hall [51], about the same level as these notes; Abraham and Marsden [1], an austere development of symplectic geometry which omits most of the details in its later chapters; Arnold [7], an intuitive book which introduces many topics but lacks proofs at times; and Siegel and Moser [81], a clearly written book with complete proofs. Of these books, Siegel and Moser is the one to read.

The classic on periodic solutions of the N-body problem is Moulton [58].

2. Equations of Celestial Mechanics

In this chapter the Hamiltonian formulation of the N-body problem is given, along with the classical integrals of energy, linear momentum, and angular momentum. Various special cases are given: the Kepler problem (also called the central force problem), the restricted three-body problem, Hill's lunar problem, and the elliptic restricted three-body problem.

2.1 Equations of the N-Body Problem

Consider N point masses moving in a Newtonian reference system, $\mathbb{R}^3$, with the only forces acting on them being their mutual gravitational attractions. Let the ith particle have position vector $\mathbf{q}_i$ and mass $m_i > 0$; then by Newton's second law and the law of gravity we have the equation of motion for the ith particle

$$m_i \ddot{\mathbf{q}}_i = \sum_{j=1}^{N} \frac{G m_i m_j (\mathbf{q}_j - \mathbf{q}_i)}{\|\mathbf{q}_i - \mathbf{q}_j\|^3} = \frac{\partial \mathbf{U}}{\partial \mathbf{q}_i}, \tag{2.1}$$

where

$$\mathbf{U} = \sum_{1 \le i < j \le N} \frac{G m_i m_j}{\|\mathbf{q}_i - \mathbf{q}_j\|}. \tag{2.2}$$

In the above, G is the universal gravitational constant, which we will take to be 1 henceforth, and $\mathbf{U}$ is the self-potential or the negative of the potential. The independent variable is time, t, and dots denote differentiation by t, so $\dot{} = d/dt$ and $\ddot{} = d^2/dt^2$. Here and throughout this book, we do not divide by zero, so the term $i = j$ is to be omitted from the sum. The system of ordinary differential equations (2.1) defines the N-body problem (the Newtonian formulation of the N-body problem).

Although Einstein's equations of relativity are thought to be the correct equations describing gravitational problems, the classical N-body problem gives an extremely accurate description of our solar system and many of the other systems of astronomy. Lunar landings and Martian probes followed trajectories of these equations.

Let $\mathbf{q} = (\mathbf{q}_1, \mathbf{q}_2, \ldots, \mathbf{q}_N) \in \mathbb{R}^{3N}$. The vector form of equation (2.1) is

$$\mathbf{M}\ddot{\mathbf{q}} - \nabla \mathbf{U}(q) = 0,$$

where $\mathbf{M} = \text{diag}(m_1, m_1, m_1, \ldots, m_N, m_N, m_N)$; the Hamiltonian formulation of the N-body problem is obtained by introducing the (linear) momentum vectors. Define $\mathbf{p} = (\mathbf{p}_1, \ldots, \mathbf{p}_N) \in \mathbb{R}^{3N}$ by $\mathbf{p} = \mathbf{M}\dot{\mathbf{q}}$ so $\mathbf{p}_i = m_i \dot{\mathbf{q}}_i$ is the momentum of the ith particle. The equations of motion become

$$\dot{\mathbf{q}} = \mathbf{H}_\mathbf{p} = \mathbf{M}^{-1}\mathbf{p}, \quad \dot{\mathbf{p}} = -\mathbf{H}_\mathbf{q} = \mathbf{U}_\mathbf{q} \tag{2.3}$$

or, in components,

$$\begin{aligned} \dot{\mathbf{q}}_i &= \frac{\partial \mathbf{H}}{\partial \mathbf{p}_i} = \mathbf{p}_i / m_i, \\ \dot{\mathbf{p}}_i &= -\frac{\partial \mathbf{H}}{\partial \mathbf{q}_i} = \frac{\partial \mathbf{U}}{\partial \mathbf{q}_i} = \sum_{j=1}^{N} \frac{m_i m_j (\mathbf{q}_j - \mathbf{q}_i)}{\|\mathbf{q}_i - \mathbf{q}_j\|^3}, \end{aligned} \tag{2.4}$$

where the Hamiltonian is

$$\mathbf{H} = \frac{1}{2}\mathbf{p}^T \mathbf{M}^{-1}\mathbf{p} - \mathbf{U} = \sum_{i=1}^{N} \frac{\|\mathbf{p}_i\|^2}{2m_i} - \mathbf{U}. \tag{2.5}$$

$\mathbf{H}$ is the total energy of the system of particles. It is an integral (i.e., a constant of the motion — see Section 5.2) of the system of equations since

$$\frac{d\mathbf{H}}{dt} = \frac{\partial \mathbf{H}}{\partial \mathbf{q}}\dot{\mathbf{q}} + \frac{\partial \mathbf{H}}{\partial \mathbf{p}}\dot{\mathbf{p}} = \frac{\partial \mathbf{H}}{\partial \mathbf{q}}\frac{\partial \mathbf{H}}{\partial \mathbf{p}} + \frac{\partial \mathbf{H}}{\partial \mathbf{p}}\left(-\frac{\partial \mathbf{H}}{\partial \mathbf{q}}\right) = 0.$$

The vectors $\mathbf{q}$ and $\mathbf{p}$ are called conjugate variables.

The N-body problem is a system of $3N$ second-order equations in the Newtonian formulation and a system of $6N$ first-order equations in the Hamiltonian formulation. A complete set of integrals for the system would comprise $6N - 1$ time-independent integrals plus one time-dependent integral. Only ten integrals are known for all N. Let

$$\mathbf{C} = m_1\mathbf{q}_1 + \ldots + m_N\mathbf{q}_N$$

be the *center of mass* of the system and

$$\mathbf{L} = \mathbf{p}_1 + \ldots + \mathbf{p}_N$$

be the *(total) linear momentum* of the system. It follows from (2.3) that

$$\dot{\mathbf{C}} = \mathbf{L}, \qquad \dot{\mathbf{L}} = 0, \qquad \ddot{\mathbf{C}} = 0, \tag{2.6}$$

and so $\mathbf{C} = \mathbf{L}_0 t + \mathbf{C}_0$, $\mathbf{L} = \mathbf{L}_0$. $\mathbf{L}_0$ and $\mathbf{C}_0$ are vector functions of the initial conditions and constants of the motion: therefore, they constitute six integrals of the motion.

Let $\mathbf{O} = \sum_1^N \mathbf{q}_i \times \mathbf{p}_i$ be *(total) angular momentum.* Since

$$\frac{d\mathbf{O}}{dt} = \sum_{i=1}^{N} (\dot{\mathbf{q}}_i \times \mathbf{p}_i + \mathbf{q}_i \times \dot{\mathbf{p}}_i)$$

$$= \sum_{i=1}^{N} (\mathbf{p}_i/m_i) \times \mathbf{p}_i + \sum_{i=1}^{N}\sum_{j=1}^{N} \mathbf{q}_i \times \frac{m_i m_j (\mathbf{q}_j - \mathbf{q}_i)}{\|\mathbf{q}_i - \mathbf{q}_j\|^3}$$

$$= 0,$$

O is a vector of integrals. Thus, there are three angular momentum integrals. Energy, center of mass, linear momentum, and angular momentum are the ten classical integrals of the N-body problem. In order to overcome some of the difficulties they cause in perturbation analysis, they are investigated in greater detail in Chapter 5.

2.2 The Kepler Problem

A special case of the two-body problem is when one body of mass M is assumed to be fixed at the origin — for example, like the sun, the body is so massive that to the first approximation it does not move. In this case, the Newtonian equation of the motion of the other body of mass m is of the form

$$m\ddot{\mathbf{q}} = -\frac{GMm\mathbf{q}}{\|\mathbf{q}\|^3},$$

or

$$\ddot{\mathbf{q}} = -\frac{\mu\mathbf{q}}{\|\mathbf{q}\|^3} = \nabla \mathbf{U}(q),$$

where $\mathbf{q} \in \mathbb{R}^3$ is the position vector of the other body in a fixed coordinate system, μ is the constant GM (G is the universal gravitational constant), and **U** is the self-potential (the negative of potential energy)

$$\mathbf{U} = \frac{\mu}{\|q\|}.$$

If we define momentum $\mathbf{p} = \dot{\mathbf{q}} \in \mathbb{R}^3$, then the Newtonian equation can be written in the Hamiltonian form

$$\dot{\mathbf{q}} = \mathbf{H}_\mathbf{p} = \mathbf{p}, \qquad \dot{\mathbf{p}} = -\mathbf{H}_\mathbf{q} = -\frac{\mu\mathbf{q}}{\|\mathbf{q}\|^3},$$

where

$$\mathbf{H} = \frac{\|\mathbf{p}\|^2}{2} - \frac{\mu}{\|\mathbf{q}\|}.$$

$\mathbf{H}$ is called the *Hamiltonian of the Kepler problem.* The Newtonian formulation is a system of three second-order scalar equations, whereas the Hamiltonian equations are six first-order scalar equations.

$\mathbf{H}$ is an integral of the motion, i.e., it is constant along solutions, since

$$\frac{d\mathbf{H}}{dt} = \frac{\partial \mathbf{H}}{\partial \mathbf{q}}\dot{\mathbf{q}} + \frac{\partial \mathbf{H}}{\partial \mathbf{p}}\dot{\mathbf{p}} = \frac{\partial \mathbf{H}}{\partial \mathbf{q}}\frac{\partial \mathbf{H}}{\partial \mathbf{p}} - \frac{\partial \mathbf{H}}{\partial \mathbf{p}}\frac{\partial \mathbf{H}}{\partial \mathbf{q}} = 0.$$

Define $\mathbf{O} = \mathbf{q} \times \mathbf{p}$, the *angular momentum.* Since

$$\dot{\mathbf{O}} = \dot{\mathbf{q}} \times \mathbf{p} + \mathbf{q} \times \dot{\mathbf{p}} = \mathbf{p} \times \mathbf{p} + \mathbf{q} \times (-\mu\mathbf{q}/\|\mathbf{q}\|^3) = 0,$$

angular momentum $\mathbf{O}$ is constant along the solutions; thus the three components of $\mathbf{O}$ are integrals. If $\mathbf{O} = 0$, then

$$\frac{d}{dt}\left(\frac{\mathbf{q}}{\|\mathbf{q}\|}\right) = \frac{(\mathbf{q} \times \dot{\mathbf{q}}) \times \mathbf{q}}{\|\mathbf{q}\|^3} = \frac{\mathbf{O} \times \mathbf{q}}{\|\mathbf{q}\|^3} = 0.$$

In the first equality above we have used the vector identity $(\mathbf{q}\cdot\mathbf{q})\dot{\mathbf{q}} - \mathbf{q}\cdot\dot{\mathbf{q}} = (\mathbf{q} \times \dot{\mathbf{q}}) \times \mathbf{q}$. Therefore, if angular momentum is zero, the motion is collinear. Letting the line of motion be one of the coordinate axes makes the problem a one degree of freedom problem and so solvable by quadrature. In this case, the integrals are elementary and one obtains simple formulas for the solutions.

If $\mathbf{O} \neq 0$, then both $\mathbf{q}$ and $\mathbf{p} = \dot{\mathbf{q}}$ are orthogonal to $\mathbf{O}$, and the motion therefore takes place in the plane orthogonal to $\mathbf{O}$ through the origin, the *invariant plane.* In this case, take one coordinate axis, say the last, to point along $\mathbf{O}$; then the motion is in a coordinate plane. The equations of motion in this coordinate plane have the same form as above, but now $\mathbf{q}, \mathbf{p} \in \mathbb{R}^2$. In the planar problem only the component of angular momentum perpendicular to the plane is nontrivial, so the problem is reduced to a problem of two degrees of freedom with one integral. Such a problem is solvable "up to quadrature." It turns out that the problem is solvable (well, almost) in terms of elementary functions. We shall solve this problem in a later section.

2.3 Restricted Problem

There are many restricted problems in celestial mechanics. The word restricted usually implies that one or more of the particles has mass zero. But *the restricted* means the restricted circular three-body problem as give in this section.

The restricted problem is a limiting case of the three-body problem when one of the masses tends to zero. A careful derivation of this problem will be given in Chapter 8. In the traditional derivation of the restricted three-body problem, one is asked to consider the motion of an infinitesimally small particle moving under the influence of the gravitational attraction of two finite particles which revolve around each other in circular orbits with uniform velocity. A full derivation will be given in Chapter 8, but for now we shall simply give the Hamiltonian. Let the two finite particles, called the primaries, have masses $\mu > 0$ and $1 - \mu > 0$. Let $q \in \mathbb{R}^2$ be the coordinate of the infinitesimal particle in a uniformly rotating coordinate system and $p \in \mathbb{R}^2$ the momentum conjugate to q. The rotating coordinate system is so chosen that the particle of mass μ is always at $(1-\mu, 0)$ and the particle of mass $1-\mu$ is at $(-\mu, 0)$. See Figure 2.1.

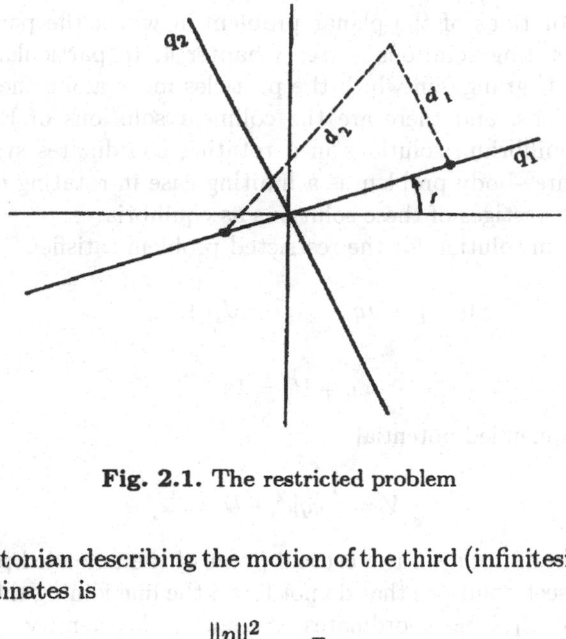

Fig. 2.1. The restricted problem

The Hamiltonian describing the motion of the third (infinitesimal) particle in these coordinates is

$$H = \frac{\|p\|^2}{2} - q^T J p - U, \tag{2.7}$$

where $q, p \in \mathbb{R}^2$ are conjugate, $J = J_2 = \begin{pmatrix} 0 & 1 \\ -1 & 0 \end{pmatrix}$, U is the self-potential

$$U = \frac{\mu}{d_1} + \frac{1-\mu}{d_2}, \tag{2.8}$$

and d_i is the distance from the infinitesimal body to the ith primary, so that

$$d_1{}^2 = (q_1 - 1 + \mu)^2 + q_2^2, \qquad d_2{}^2 = (q_1 + \mu)^2 + q_2^2.$$

The equations of motion are

$$\dot{q} = H_p = Jq + p,$$

$$\dot{p} = -H_q = Jp + U_q.$$

The term $q^T Jp$ in the Hamiltonian H is due to the fact that the coordinate system is not a Newtonian system, but instead a rotating coordinate system. It gives rise to the so-called Coriolis forces in the equations of motion. The line joining the masses is known as the *line of syzygy.*

The proper definition of the restricted three-body problem is the system of differential equations defined by the Hamiltonian above. It is a two degrees of freedom problem that seems simple but has defied integration; it has given rise to an extensive body of research and will be treated in detail in Chapter 8.

The full three-body problem has no equilibrium points, but as we will see, there are solutions of the planar problem in which the particles move on uniformly rotating solutions — see Chapter 4. In particular, there are the solutions of Lagrange, in which the particles move along the equilateral triangular solutions, and there are the collinear solutions of Euler. These solutions are equilibrium solutions in a rotating coordinates system. Since the restricted three-body problem is a limiting case in rotating coordinates, we expect to see vestiges of these solutions as equilibria.

An equilibrium solution for the restricted problem satisfies

$$0 = p + Jq, \qquad 0 = Jp + U_q,$$

which implies

$$0 = q + U_q = V_q,$$

where V is the amended potential

$$V = \frac{1}{2}\|q\|^2 + U.$$

Thus an equilibrium solution is a critical point of the amended potential.

First, let us seek solutions that do not lie on the line joining the primaries. Use the distances d_1, d_2 as coordinates. We obtain the identity

$$q_1^2 + q_2^2 = \mu d_1^2 + (1-\mu)d_2^2 - \mu(1-\mu),$$

so V can be written in terms of the distances d_1 and d_2. The equation $V_q = 0$ becomes in these variables

$$\mu d_1 - \frac{\mu}{d_1^2} = 0, \qquad (1-\mu)d_2 - \frac{(1-\mu)}{d_2^2} = 0,$$

which clearly has the unique solution $d_1 = d_2 = 1$. This solution lies at the vertex of an equilateral triangle whose base is the line segment joining the two primaries. Since there are two orientations, there are two such equilibria solutions; one in the lower half-plane is denoted by $\mathcal{L}_4$, and one in the upper half-plane is denoted by $\mathcal{L}_5$. These solutions are attributed to Lagrange also.

Lagrange thought that these solutions had no astronomical significance, but he was wrong: in the twentieth century, such a system was discovered in our own neighborhood. Consider a line segment connecting the sun and Jupiter as the base of an equilateral triangle in the sun-Jupiter plane. One group of about fifteen asteroids, called the Trojans, is found at $\mathcal{L}_4$ and another group of about fifteen, called the Greeks, at $\mathcal{L}_5$.

Now consider equilibria along the line of the primaries where $q_2 = 0$. In this case the amended potential is a function of q_2, which we shall denote by q for the present, and has the form

$$V = \frac{1}{2}q^2 \pm \frac{\mu}{(q-1+\mu)} \pm \frac{(1-\mu)}{(q+\mu)}.$$

In the above, one takes the signs so that each term is positive. There are three cases: (i) when $q < -\mu$, the signs are $-$ and $-$; (ii) when $-\mu < q < 1-\mu$, the signs are $-$ and $+$; and (iii) when $1-\mu < q$, the signs are $+$ and $+$. Clearly $V \to \infty$ as $q \to \pm\infty$ or as $q \to -\mu$ or as $q \to 1-\mu$, so V has at least one critical point on each of these three intervals. Also note that

$$\frac{d^2V}{dq^2} = 1 \pm \frac{\mu}{(q-1+\mu)^3} \pm \frac{(1-\mu)}{(q+\mu)^3},$$

where the signs are again taken so that each term is positive, so V is a convex function having precisely one critical point in each of these intervals, or three critical points. A sketch of the graph of V is given in Figure 2.2. These three

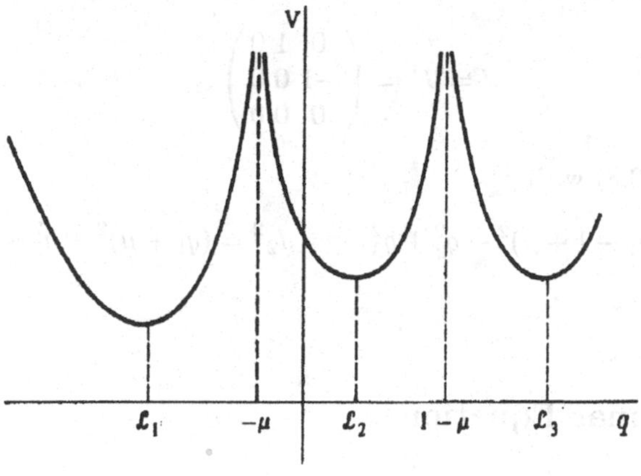

Fig. 2.2. The amended potential

collinear equilibria are attributed to Euler and are denoted by $\mathcal{L}_1, \mathcal{L}_2, \mathcal{L}_3$, as shown in Figure 2.3. In classical celestial mechanics literature, these equilibrium points are called *libration points*, hence the use of the symbol $\mathcal{L}$.

One can consider the spatial restricted three-body problem also. In that case the Hamiltonian is of the same form as (2.7) except now $p = (p_1, p_2, p_3) \in$

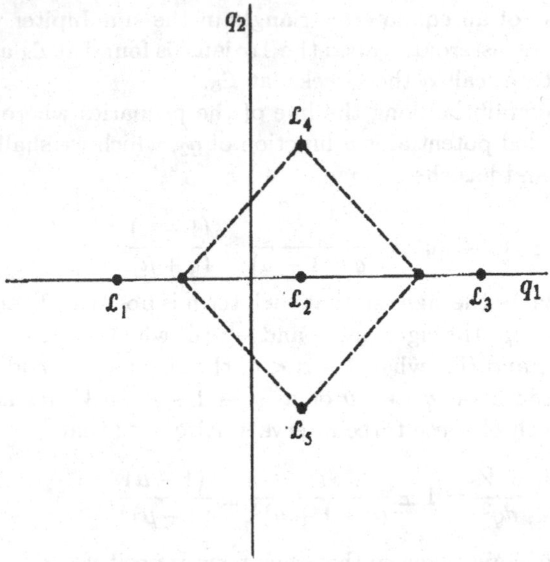

Fig. 2.3. The equilibrium points of the restricted problem

$\mathbb{R}^3$, $q = (q_1, q_2, q_3) \in \mathbb{R}^3$,

$$J = J^* = \begin{pmatrix} 0 & 1 & 0 \\ -1 & 0 & 0 \\ 0 & 0 & 0 \end{pmatrix},$$

and U is as in (2.8) with

$$d_1{}^2 = (q_1 - 1 + \mu)^2 + q_2^2 + q_3^2, \qquad d_2{}^2 = (q_1 + \mu)^2 + q_2^2 + q_3^2.$$

2.4 Hill's Lunar Equations

In a popular description of Hill's lunar equations, one is asked to consider the motion of an infinitesimal body (the moon) which is attracted to a body (the earth) fixed at the origin. The infinitesimal body moves in a rotating coordinate system: the system rotates so that the positive x-axis points to an infinite body (the sun) infinitely far away. The ratio of the two infinite quantities is taken so that the gravitational attraction of the sun on the moon is finite. A careful derivation of these equations will be given in Chapter 11, along with a discussion of why Hill's definition of the "main problem" is important in lunar theory. For now we will simply give its Hamiltonian:

$$H = \frac{1}{2}\|p\|^2 - q^T Jp - \frac{1}{\|q\|} - q_1^2 + \frac{1}{2}q_2^2. \tag{2.9}$$

2.5 Elliptic Restricted Problem

Previously we gave the Hamiltonian of the circular restricted three-body problem. In that problem, it is assumed that the two primaries move on circular orbits of the two-body problem. If one assumes instead that the primaries move on elliptical orbits of the two-body problem, then one has the elliptical restricted problem. It is no longer autonomous (time-independent) but rather periodic in time. It also contains a parameter e, denoting the eccentricity of the primaries' orbit. In Chapter 12, we give a detailed derivation of these equations and the special coordinates used to study this problem; for now we will just give the Hamiltonian.

The Hamiltonian governing the motion of the third (infinitesimal) particle in the elliptic restricted problem is

$$H = \frac{\|p\|^2}{2} - q^T Jp - r(t)U + \left(\frac{1 - r(t)}{2}\right) q^T q,$$

where $q, p \in \mathbb{R}^2$ are conjugate, J is J_2, U is the self-potential

$$U = \frac{\mu}{d_1} + \frac{1-\mu}{d_2},$$

d_i is the distance from the infinitesimal body to the ith primary, or

$$d_1^2 = (q_1 - 1 + \mu)^2 + q_2^2, \qquad d_2^2 = (q_1 + \mu)^2 + q_2^2,$$

and $r(t) = 1/(1 + e\cos t)$. These are special coordinates that keep the primaries at a fixed position.

In all the other examples of this chapter, the Hamiltonian is independent of time t, so it is a constant of the motion or an integral. However, in the elliptic case, the Hamiltonian is not independent of time and so is not an integral.

2.6 Problems

1 Show that the Hill's lunar equations have two equilibrium points on the q_1 axis. Analyze the linearized equations at these equilibrium points.

2 Show that the elliptic restricted three-body problem has five equilibrium points — two at the vertices of an equilateral triangle and three collinear.

3 Show that there are no new equilibrium points in:

- the spatial circular restricted three-body problem,
- the spatial elliptic restricted three-body problem,
- the spatial Hill's lunar problem.

3. Hamiltonian Systems

In this chapter we review some of the basic concepts of the Hamiltonian formalism and symplectic geometry, but do not give a complete mathematical development of all of them.

3.1 Hamiltonian Systems

Hamiltonian systems have already been introduced by example, when we wrote the equations of motion of the Kepler problem, the N-body problem, etc. as Hamiltonian systems. Before proceeding, some basic facts about such systems will be summarized; more details and proofs can be found in Meyer and Hall [51]. The Hamiltonian formalism is the natural mathematical structure in which to develop the theory of conservative mechanical systems, especially the equations of celestial mechanics.

A general Hamiltonian system is a system of $2n$ ordinary differential equations of the form

$$\dot{u} = H_v, \qquad \dot{v} = -H_u \tag{3.1}$$

or, in components,

$$\dot{u}_i = \frac{\partial H(t,u,v)}{\partial v_i}, \qquad \dot{v}_i = -\frac{\partial H(t,u,v)}{\partial u_i}, \tag{3.2}$$

where $H = H(t,u,v)$, called the *Hamiltonian*, is a smooth real-valued function defined for $(t,u,v) \in \mathcal{O}$, an open set in $\mathbb{R}^1 \times \mathbb{R}^n \times \mathbb{R}^n$. The vectors $u = (u_1, \ldots, u_n)$ and $v = (v_1, \ldots, v_n)$ are traditionally called the position and momentum vectors, respectively, and t is called time. The variables u and v are said to be *conjugate* variables: v is conjugate to u, and u is conjugate to v. The integer n is the *number of degrees of freedom* of the system.

For the general discussion, introduce the $2n$ vector

$$z = \begin{pmatrix} u \\ v \end{pmatrix}$$

and the $2n \times 2n$ skew symmetric matrix

$$J = J_{2n} = \begin{pmatrix} 0 & I \\ -I & 0 \end{pmatrix}$$

where 0 is the $n \times n$ zero matrix and I is the $n \times n$ identity matrix. Usually we will use J without subscript — the size will be determined by the context. With this notation, equation (3.1) becomes

$$\dot{z} = J\nabla_z H(t, z). \tag{3.3}$$

One of the basic results from the general theory of ordinary differential equations is the existence and uniqueness theorem. This theorem states that for each $(\tau, \zeta) \in \mathcal{O}$, there is a unique solution $z = \phi(t, \tau, \zeta)$ of (3.3) defined for t near τ satisfying the initial condition $\phi(\tau, \tau, \zeta) = \zeta$. The function $\phi(t, \tau, \zeta)$ is smooth in all its displayed arguments, and so analytic if the equations are analytic. We call $\phi(t, \tau, \zeta)$ the *general solution.*

In the special case when H is independent of t, so that we have $H : \mathcal{O} \to \mathbb{R}^1$ where $\mathcal{O}$ is some open set in $\mathbb{R}^{2n}$, the differential equations (3.3) are *autonomous* and the Hamiltonian system is called *conservative.* In this case, the identity $\phi(t - \tau, 0, \zeta) = \phi(t, \tau, \zeta)$ holds, since both sides satisfy the equations (3.3) and the same initial conditions. In this case, the τ dependence is usually dropped and only $\phi(t, \zeta)$ is considered, where $\phi(t, \zeta)$ is the solution of (3.3) satisfying $\phi(0, \zeta) = \zeta$, and we say that the equation defines a *local flow*, i.e.,

$$\phi(t, \phi(s, \zeta)) = \phi(t + s, \zeta) \tag{3.4}$$

for all t, s, and ζ for which all the quantities in this formula are defined, in particular, for t and s small. If the solutions are defined for all t, then the above holds for all t and s and ϕ is a *flow.*

3.2 Symplectic Coordinates

The form of Hamilton's equations is very special. The special form is not preserved by an arbitrary change of variables, so the change of variables that does preserve that special form is very important in the theory. The classical subject of celestial mechanics is replete with special coordinate systems which bear the names of some of the greatest mathematicians.

A $2n \times 2n$ matrix S is called *symplectic* if it satisfies

$$S^T J S = J. \tag{3.5}$$

The set of all $2n \times 2n$ symplectic matrices forms a group called the *symplectic group* and is denoted by Sp_n or $Sp(2n, \mathbb{R})$. The determinant of a symplectic matrix is $+1$ and the transpose of a symplectic matrix is symplectic also.

Let $T : \mathcal{O} \to \mathbb{R}^{2n} : (t, z) \to Z = T(t, z)$ be a smooth function with $\mathcal{O}$ some open set in $\mathbb{R}^{2n+1}$. T is called a *symplectic map* (or *transformation* or *function*, etc.) if its Jacobian $\partial T/\partial z$ is a symplectic matrix for all $(t,z) \in \mathcal{O}$. The composition of two symplectic maps is symplectic and the inverse of a symplectic map is symplectic. (The inverse function theorem implies that a symplectic map is locally invertible.) Since the determinant of a symplectic matrix is $+1$, a symplectic transformation is orientation- and volume-preserving.

If the transformation $z \to Z = T(t,z)$ is considered a change of variables, then one calls Z the *symplectic* or *canonical coordinates.* Consider a nonlinear Hamiltonian system $\dot{z} = J\nabla_z H(t,z)$ and make the change of variables from z to Z by $Z = T(t,z) = Z(t,z)$ with inverse $z = T^{-1}(t,Z) = z(t,Z)$. Then the Hamiltonian $H(t,z)$ transforms to the function $K(t,Z) = H(t,z(t,Z))$. Later we will abuse notation and write $H(t,Z)$ instead of introducing a new symbol, but now we will be careful to distinguish H and K. It can be shown [51] that the equation (3.3) transforms to

$$\dot{Z} = J\nabla_Z K(t,Z) + J\nabla_Z R(t,Z),$$

where R is defined by the formula

$$\frac{\partial T}{\partial t}(t,z)\Big|_{z=z(t,Z)} = J\nabla_Z R(t,Z).$$

R is called the *remainder function.* Note that if the transformation is independent of time t, then the remainder is zero. Therefore, in the new coordinates, the equation is Hamiltonian with Hamiltonian $K(t,Z) + R(t,Z)$. (Here we assume that $\mathcal{O}$ is simply connected so that a closed 1-form is exact; otherwise, R is only locally defined.)

Proposition 3.2.1. *A symplectic change of variables takes a Hamiltonian system of equations into a Hamiltonian system. If a change of variables preserves the Hamiltonian form of all Hamiltonian equations, then it is symplectic.*

Let $\phi(t,\tau,\zeta)$ be the general solution of (3.3), so $\phi(\tau,\tau,\zeta) = \zeta$, and let $X(t,\tau,\zeta)$ be the Jacobian of ϕ with respect to ζ, that is, $X(t,\tau,\zeta) = (\partial\phi/\partial\zeta)(t,\tau,\zeta)$. $X(t,\tau,\zeta)$ is called the *monodromy matrix.* Substituting ϕ into (3.3) and differentiating with respect to ζ, we get

$$\dot{X} = JS(t,\tau,\zeta)X, \qquad S(t,\tau,\zeta) = \frac{\partial^2 H}{\partial z^2}(t,\phi(t,\tau,\zeta)).$$

This equation is called the *variational equation* and is a linear Hamiltonian system. Differentiating the identity $\phi(\tau,\tau,\zeta) = \zeta$ with respect to ζ gives $X(\tau,\tau,\zeta) = I$, the $2n \times 2n$ identity matrix, so X is a fundamental matrix solution of the variational equation. (Recall that a fundamental matrix solution $X(t,\tau)$ is a square matrix solution of a linear equation that satisfies $X(\tau,\tau) = I$.) Therefore, X is symplectic by the following general result.

Proposition 3.2.2. *The fundamental matrix solution of a linear Hamiltonian system is symplectic for all t.*

This means that the flow defined by an autonomous Hamiltonian system is volume-preserving, so in particular there cannot be an asymptotically stable equilibrium point, periodic solution, etc. This makes the stability theory of Hamiltonian systems both difficult and interesting.

In the conservative case, the equations define a flow $\phi(t,\zeta)$ and the above implies that the map $\zeta \longrightarrow \phi(t,\zeta)$ is symplectic where defined. Such a flow will be called a *(local) symplectic flow.* The converse is partially true.

Proposition 3.2.3. *If $\phi(t,\zeta)$ is a local symplectic flow for t small and $\zeta \in \mathcal{O}' \subset \mathbb{R}^{2n}$ where $\mathcal{O}'$ is simply connected, then there is a smooth function $H : \mathcal{O}' \longrightarrow \mathbb{R}$ such that $\phi(t,\zeta)$ is the general solution of $\dot{z} = J\nabla_z H(z)$.*

The definition of a symplectic transformation is easy enough to check a posteriori, but it is difficult to use this definition to generate a symplectic transformation with the desired properties. Here are some results which help in constructing symplectic transformations.

The differential form

$$\Omega = \frac{1}{2}\sum_{i=1}^{2n}\sum_{j=1}^{2n} J_{ij}dz^i \wedge dz^j = \sum_{i=1}^{n} dz^i \wedge dz^{i+n} = \sum_{i=1}^{n} du^i \wedge dv^i = du \wedge dv$$

is the *standard symplectic form.*

Proposition 3.2.4. *A transformation $z \longrightarrow \zeta$ is symplectic if and only if it preserves the standard symplectic form, i.e.,*

$$\sum_{i=1}^{n} dz^i \wedge dz^{i+n} = \sum_{i=1}^{n} d\zeta^i \wedge d\zeta^{i+n}.$$

See [1, 51].

For example, the change from rectangular coordinates u, v to polar coordinates r, θ is not symplectic since $du \wedge dv = r\,dr \wedge d\theta$. But by defining $I = r^2/2 = (u^2 + v^2)/2$, we have $du \wedge dv = dI \wedge d\theta$, so the transformation $u, v \longrightarrow I, \theta$ is symplectic and I, θ are symplectic coordinates. They are known as *action-angle variables.*

3.3 Generating Functions

Use classical notation $z = (u, v)$ so that the standard symplectic form is

$$\Omega = \sum_{i=1}^{n} du^i \wedge dv^i = du \wedge dv.$$

Let $\bar{u} = \bar{u}(u, v), \bar{v} = \bar{v}(u, v)$ be a change of variables, and assume that the functions $\bar{u}$ and $\bar{v}$ are defined in a ball in $\mathbb{R}^{2n}$. This change of variables is symplectic if and only if

$$du \wedge dv = d\bar{u} \wedge d\bar{v}.$$

This is equivalent to $d(u\,dv - \bar{u}\,d\bar{v}) = 0$ or that $\sigma_1 = u\,dv - \bar{u}\,d\bar{v}$ is closed. σ_1 is closed if and only if $\sigma_2 = \sigma_1 + d(\bar{u}\bar{v}) = udv + \bar{v}d\bar{u}$ is closed. In a similar manner, the change of variables $\bar{u} = \bar{u}(u,v), \bar{v} = \bar{v}(u,v)$ is symplectic if and only if any one of the following forms is closed:

$$\begin{aligned} \sigma_1 &= u\,dv - \bar{u}\,d\bar{v}, \qquad & \sigma_2 &= u\,dv + \bar{v}\,d\bar{u}, \\ \sigma_3 &= v\,du - \bar{v}\,d\bar{u}, \qquad & \sigma_4 &= v\,du + \bar{u}\,d\bar{v}. \end{aligned} \tag{3.6}$$

Since the functions $\bar{u}$ and $\bar{v}$ are defined in a ball, closed forms are exact by Poincaré's lemma; so the change of variables is symplectic if and only if one of the functions S_1, S_2, S_3, S_4 exists and satisfies one of

$$\begin{aligned} dS_1(v,\bar{v}) &= \sigma_1, \qquad & dS_2(v,\bar{u}) &= \sigma_2, \\ dS_3(u,\bar{u}) &= \sigma_3, \qquad & dS_4(u,\bar{v}) &= \sigma_4. \end{aligned}$$

In the above formulas, there is an implied summation over the components.

These statements give an easy way to construct a symplectic change of variables. Assume that there exists a function $S_1(v,\bar{v})$ such that $dS_1 = \sigma_1$; then

$$dS_1 = \frac{\partial S_1}{\partial v} dv + \frac{\partial S_1}{\partial \bar{v}} d\bar{v} = u\,dv - \bar{u}\,d\bar{v}.$$

So if

$$u = \frac{\partial S_1}{\partial v}(v,\bar{v}), \qquad \bar{u} = -\frac{\partial S_1}{\partial \bar{v}}(v,\bar{v}) \tag{3.7}$$

defines a change of variables from (u,v) to $(\bar{u},\bar{v})$, then it is symplectic. By the implicit function theorem, the first equations in (3.7) is solvable for $\bar{v}$ as a function of u and v when the Hessian of S_1 is nonsingular. Replacing $\bar{v} = \bar{v}(u,v)$ into the second equation gives $\bar{u} = \bar{u}(u,v)$ and this defines a symplectic change of coordinates. In a similar manner we have

Proposition 3.3.1. *The following define a local symplectic change of variables:*

$$\begin{aligned}
u &= \frac{\partial S_1}{\partial v}(v,\bar{v}), \quad \bar{u} = -\frac{\partial S_1}{\partial \bar{v}}(v,\bar{v}) \text{ when } \frac{\partial^2 S_1}{\partial v \partial \bar{v}} \text{ is nonsingular;} \\
u &= \frac{\partial S_2}{\partial v}(v,\bar{u}), \quad \bar{v} = \frac{\partial S_2}{\partial \bar{u}}(v,\bar{u}) \text{ when } \frac{\partial^2 S_2}{\partial v \partial \bar{u}} \text{ is nonsingular;} \\
v &= \frac{\partial S_3}{\partial u}(u,\bar{u}), \quad \bar{v} = -\frac{\partial S_3}{\partial \bar{u}}(u,\bar{u}) \text{ when } \frac{\partial^2 S_3}{\partial u \partial \bar{u}} \text{ is nonsingular;} \\
v &= \frac{\partial S_4}{\partial u}(u,\bar{v}), \quad \bar{u} = \frac{\partial S_4}{\partial \bar{v}}(u,\bar{v}) \text{ when } \frac{\partial^2 S_4}{\partial u \partial \bar{v}} \text{ is nonsingular.}
\end{aligned} \tag{3.8}$$

(Remark: In the above the partial derivatives are to be column vectors.) The functions S_i are called *generating functions.* For example, if $S_2(v, \bar{u}) = v\bar{u}$, then the identity transformation $\bar{u} = u, \bar{v} = v$ is symplectic, and if $S_1(v, \bar{v}) = v\bar{v}$, then the switching of variables $\bar{u} = -u, \bar{v} = v$ is symplectic.

If we are given a point transformation $\bar{u} = f(u)$ with $\partial f / \partial u$ invertible, then the transformation can be extended to a symplectic transformation by defining $S_4(u, \bar{v}) = f(u)^T \bar{v}$ and

$$v = \frac{\partial f}{\partial u}(u)\bar{v}, \qquad \bar{u} = f(u).$$

These transformations are called *Mathieu transformations.* The particular case in which $f(u) = A^T u$, where A is a nonsingular $n \times n$ matrix gives the linear symplectic transformation whose matrix is

$$\begin{pmatrix} A^T & 0 \\ 0 & A^{-1} \end{pmatrix}.$$

3.4 Rotating Coordinates

Let $J = J_2 = \begin{pmatrix} 0 & 1 \\ -1 & 0 \end{pmatrix}$, $\exp(\omega J t) = \begin{pmatrix} \cos \omega t & \sin \omega t \\ -\sin \omega t & \cos \omega t \end{pmatrix}$ be 2×2 matrices, and consider the planar N-body problem; so the vectors $\mathbf{q}_i, \mathbf{p}_i$ are 2-vectors. Introduce a set of coordinates that uniformly rotate with frequency ω by

$$q_i = \exp(\omega J t)\mathbf{q}_i, \qquad p_i = \exp(\omega J t)\mathbf{p}_i.$$

Since J is skew-symmetric, $\exp(\omega J t)$ is orthogonal for all t, so the change of variables is symplectic. The remainder function is $-\sum \omega q_i^T J p_i$, and so the Hamiltonian of the N-body problem in rotating coordinates is

$$H = \sum_{i=1}^{N} \frac{\|p_i\|^2}{2m_i} - \sum_{i=1}^{N} \omega q_i^T J p_i - \sum_{1 \le i,j \le N} \frac{m_i m_j}{\|q_i - q_j\|}.$$

The remainder term gives rise to extra terms in the equations of motion which are sometimes called Coriolis forces.

The equations of motion are

$$\begin{aligned} \dot{q}_i &= \frac{\partial H}{\partial p_i} = \frac{p_i}{m_i} + \omega J q_i, \\ \dot{p}_i &= -\frac{\partial H}{\partial q_i} = \omega J p_i + \sum_{j=1}^{N} \frac{m_i m_j (q_j - q_i)}{\|q_j - q_j\|^3}. \end{aligned} \tag{3.9}$$

Usually, we will take $\omega = 1$ in our discussions. A great deal of the discussion of the N-body problem in the following chapters will be in rotating coordinates. Fixed coordinates will be in a boldface font and rotating coordinates will be in a regular font.

The Kepler problem in rotating coordinates is

$$H = \frac{\|p\|^2}{2} - q^T J p - \frac{\mu}{\|q\|}.$$

We can use rotating coordinates for the spatial problem also. Let

$$J = J^* = \begin{pmatrix} 0 & 1 & 0 \\ -1 & 0 & 0 \\ 0 & 0 & 0 \end{pmatrix}, \qquad \exp(\omega J t) = \begin{pmatrix} \cos\omega t & \sin\omega t & 0 \\ -\sin\omega t & \cos\omega t & 0 \\ 0 & 0 & 1 \end{pmatrix}$$

be 3×3 matrices, and consider the spatial N-body problem; so the vectors $\mathbf{q}_i, \mathbf{p}_i$ are 3-vectors. Introduce a set of coordinates that uniformly rotate about the $\mathbf{q}_3$ axis with frequency ω by

$$q_i = \exp(\omega J t)\mathbf{q}_i, \qquad p_i = \exp(\omega J t)\mathbf{p}_i.$$

The Hamiltonian has the same form as for the planar problem with the new definition of J.

3.5 Jacobi Coordinates

Jacobi coordinates are ideal for the problems considered in this book for several reasons. First, one coordinate locates the center of mass of the system; thus it can be set to zero and ignored in subsequent considerations. This accomplishes the first reduction of the dimension of the problem — see Chapter 5 on reduction. Second, another coordinate is the vector from one particle to another, so it can be easily scaled in the problem in which two of the particles are close — see Chapter 9 on lunar orbits. Third, another coordinate is the vector to one particle from the center of mass of the other particles, and it can be easily scaled in the problem in which one particle is far from the others — see Chapter 10 on comet orbits. Last, the Hamiltonian and angular momentum are relatively simple in these coordinates.

Because of the nature of the problems considered in this book, it is necessary to discuss the N- and $(N+1)$-body problems simultaneously and in fixed and rotating coordinates. For later applications, it is convenient to consider the $(N+1)$-body problem here and to index the masses, position vectors, and momentum vectors from 0 to N. Jacobi coordinates to be introduced now work in fixed and rotating coordinates. Since we will use rotating coordinates more often, this discussion will be for rotating coordinates. The treatment for fixed coordinates is the same.

Let $q_0, q_1, \dots, q_N, p_0, \dots, p_N$ be the rotating coordinates used in the previous sections. Set $g_0 = q_0$ and $\mu_0 = m_0$. Define a sequence of point transformations by

$$T_k: \quad \begin{aligned} x_k &= q_k - g_{k-1}, \\ g_k &= \frac{1}{\mu_k}(m_k q_k + \mu_{k-1} g_{k-1}), \\ \mu_k &= m_k + \mu_{k-1} \end{aligned} \tag{3.10}$$

for $k = 1, \dots, N$. Thus μ_k is the total mass and g_k is the center of mass of the particles with index $0, 1, \dots, k$. The vector x_k is the position of the kth particle relative to the center of mass of the previous particles. Consider T_k as a change of coordinates from $g_{k-1}, x_1, \dots, x_{k-1}, q_k, \dots, q_N$ to $g_k, x_1, \dots, x_k, q_{k+1}, \dots, q_N$ or simply from g_{k-1}, q_k to g_k, x_k. The inverse of T_k is

$$T_k^{-1}: \quad \begin{aligned} q_k &= \frac{\mu_{k-1}}{\mu_k} x_k + g_k, \\ g_{k-1} &= -\frac{m_k}{\mu_k} x_k + g_k. \end{aligned} \tag{3.11}$$

In order to make the linear symplectic extension of T_k (the Mathieu transformation), define $G_0 = p_0$ and

$$Q_k: \quad \begin{aligned} y_k &= \frac{\mu_{k-1}}{\mu_k} p_k - \frac{m_k}{\mu_k} G_{k-1}, \\ G_k &= p_k + G_{k-1} \end{aligned} \tag{3.12}$$

and

$$Q_k^{-1}: \quad \begin{aligned} p_k &= y_k + \frac{m_k}{\mu_k} G_k, \\ G_{k-1} &= -y_k + \frac{\mu_{k-1}}{\mu_k} G_k. \end{aligned} \tag{3.13}$$

If we denote the coefficient matrix in (3.10) by A, then the coefficient matrices in (3.11), (3.12), and (3.13) are A^{-1}, A^{T-1}, and A^T, respectively, and the pair T_k, Q_k is a symplectic change of coordinates.

An easy calculation yields

$$g_{k-1}^T J G_{k-1} + q_k^T J p_k = g_k^T J G_k + x_k^T J y_k \tag{3.14}$$

and

$$\frac{1}{2\mu_{k-1}} \|G_{k-1}\|^2 + \frac{1}{2m_k} \|p_k\|^2 = \frac{1}{2\mu_k} |G_k\|^2 + \frac{1}{2M_k} \|y_k\|^2, \tag{3.15}$$

where $M_k = m_k \mu_{k-1} / \mu_k$.

Since each transformation T_k, Q_k is symplectic for $k = 1, \dots, N$, the composition of them is symplectic and so the change of variables from $q_0, \dots, q_N$, $p_0, \dots, p_N$ to $g_N, x_1, \dots, x_N, G_N, y_1, \dots, y_N$ is symplectic. A simple induction on (3.14) and (3.15) shows that kinetic energy is

$$K = \sum_{i=0}^{N} \frac{1}{2m_i} \|p_i\|^2 = \frac{1}{2\mu_N} \|G_n\|^2 + \sum_{i=1}^{N} \frac{1}{2M_i} \|y_i\|^2 \tag{3.16}$$

and total angular momentum O is

$$O = \sum_{i=0}^{N} q_i^T J p_i = g_N^T J G_N + \sum_{i=1}^{N} x_i^T J y_i. \tag{3.17}$$

Also g_N is the center of mass of the system and G_N is total linear momentum.

This inductive definition does not lend itself to simple formulas for the x's and y's in terms of the q's and p's, but we require only a few special properties of this representation. First note from (3.10) that

$$x_1 = q_1 - q_0. \tag{3.18}$$

We claim that

$$q_0 = g_k - \sum_{l=1}^{k} \frac{m_l}{\mu_l} x_l \quad \text{for } k = 1, \dots, N. \tag{3.19}$$

Equation (3.19) is true when $k = 1$ since (3.11) gives $g_0 = (-m_1/\mu_1)x_1 + g_1$ and $g_0 = q_0$. Assume (3.19) for $k - 1$. So $q_0 = g_{k-1} - \sum_{l=1}^{k-1} (m_l/\mu_l) x_l$, but by (3.11) again, we have $g_{k-1} = (-m_k/\mu_k)x_k + g_k$, and these two formulas yield (3.19).

Finally, we claim that

$$q_j - q_i = x_j + \sum_{l=1}^{j-1} a_{jil} x_l \quad \text{for} \quad 0 \le i < j \le N, \tag{3.20}$$

where a_{jil} are constants. We prove (3.20) by induction on j. For $j = 1$, this is just (3.18). Now assume (3.20) for $j - 1$. We need only consider $j > i$ and so

$$q_j - q_i = (q_j - q_0) - (q_i - q_0). \tag{3.21}$$

By (3.19), we have $q_0 = g_{j-1} - \sum_{l=1}^{j-1} (m_l/\mu_l) x_l$ and by (3.10), $q_j = x_j + g_{j-1}$. Since $i < j$, the induction hypothesis yields $q_i - q_0 = x_i + \sum_{l=1}^{i-1} a_{i0l} x_l$. Substituting these last three relations into (3.21) yields (3.20).

Let $d_{ji} = q_j - q_i = x_j + \sum_{l=1}^{j-1} a_{jil} x_l$. The Hamiltonian of the (N+1)-body problem in rotating Jacobi coordinates becomes

$$H = \frac{1}{2\mu_N}\|G_N\|^2 + \sum_{i=1}^{N} \frac{1}{2M_i}\|y_i\|^2 - g_N^T J G_N$$
$$- \sum_{i=1}^{N} x_i^T J y_i - \sum_{0 \le i < j \le N} \frac{m_i m_j}{\|d_{ij}\|}. \tag{3.22}$$

By (3.20), the last term in (3.22), the potential energy, is independent of g_N, and so the equations for g_N and G_N are

$$\dot{g}_N = J g_N + \frac{1}{\mu_N} G_N, \qquad \dot{G}_N = J G_N. \tag{3.23}$$

Thus the Hamiltonian of the N-body problem in rotating Jacobi coordinates on the invariant space where $g_N = G_N = 0$ is

$$H = \sum_{i=1}^{N} \left(\frac{1}{2M_i}\|y_i\|^2 - x_i^T J y_i \right) - \sum_{0 \le i < j \le N} \frac{m_i m_j}{\|d_{ji}\|}. \tag{3.24}$$

In a similar manner, the Hamiltonian of the N-body problem in fixed Jacobi coordinates $\mathbf{g}, \mathbf{x}_1, \ldots, \mathbf{x}_N, \mathbf{G}_N, \mathbf{y}_1, \ldots, \mathbf{y}_N$ on the invariant space where $\mathbf{g}_N = \mathbf{G}_N = 0$ is

$$H = \sum_{i=1}^{N} \frac{1}{2M_i}\|\mathbf{y}_i\|^2 - \sum_{0 \le i < j \le N} \frac{m_i m_j}{\|d_{ji}\|}. \tag{3.25}$$

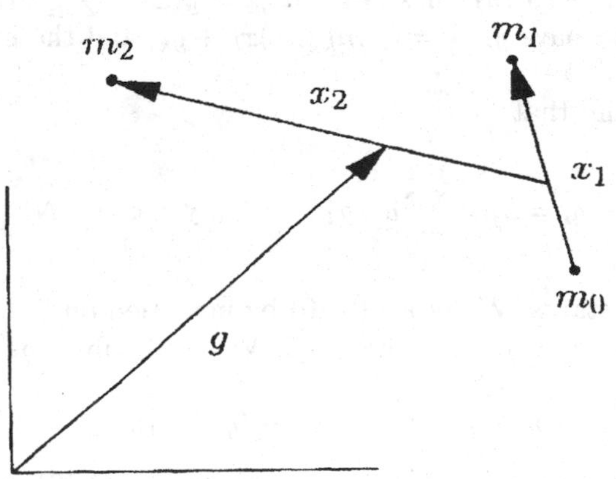

Fig. 3.1. Three-body problem in Jacobi coordinates

Consider the two-body problem in fixed Jacobi coordinates. When $N = 1$ and $g_1 = G_1 = 0$, the Hamiltonian takes the simple form

$$H = \frac{\|\mathbf{y}\|^2}{2M} - \frac{m_1 m_2}{\|\mathbf{x}\|},$$

where $\mathbf{y} = \mathbf{y}_1$, $\mathbf{x} = \mathbf{x}_1 = \mathbf{q}_1 - \mathbf{q}_0$, $M = m_0 m_1/(m_0 + m_1)$. This is just the Kepler problem, so in Jacobi coordinates the two-body problem is just the Kepler problem. This says that the motion of the moon, when viewed from the earth, is as if the earth is a fixed body and the moon is attracted to the earth as a central force. (Is the earth the center of the universe?)

Now consider the three-body problem in fixed Jacobi coordinates. In the three-body problems, the distances between the bodies and hence the potential are not too complicated in Jacobi coordinates. Moreover, the Hamiltonian of the three-body problem in Jacobi coordinates will be transformed to polar coordinates in the next section, which will be used in Chapter 7 to understand reduction of the three-body problem and to establish the existence of periodic solutions for two small masses (Poincaré's periodic solutions of the first kind).

Let

$$M_1 = \frac{m_0 m_1}{m_0 + m_1}, \ M_2 = \frac{m_2(m_0 + m_1)}{m_0 + m_1 + m_2},$$

$$\alpha_0 = \frac{m_0}{m_0 + m_1}, \ \alpha_1 = \frac{m_1}{m_0 + m_1}.$$

Then the Hamiltonian of the three-body problem with center of mass fixed at the origin and zero linear momentum in Jacobi coordinates is

$$H = \frac{\|\mathbf{y}_1\|^2}{2M_1} + \frac{\|\mathbf{y}_2\|^2}{2M_2} - \frac{m_0 m_1}{\|\mathbf{x}_1\|} - \frac{m_1 m_2}{\|\mathbf{x}_2 - \alpha_0 \mathbf{x}_1\|} - \frac{m_2 m_0}{\|\mathbf{x}_2 + \alpha_1 \mathbf{x}_1\|}.$$

See Figure 3.1. Sometimes one numbers the N bodies from 1 to N. In this case all the subscripts in the above except the subscripts of the α's are increased by 1, which looks nicer to some people.

3.6 Action-Angle and Polar Coordinates

There are two forms of polar coordinates in symplectic geometry. First let $\mathbf{q}, \mathbf{p}$ be rectangular coordinates in the plane, so $\mathbf{q} \in \mathbb{R}^1$ and $\mathbf{p} \in \mathbb{R}^1$. That is, we are considering a one degree of freedom problem. The change from the rectangular coordinates $\mathbf{q}, \mathbf{p}$ to the usual polar coordinates r, θ is not symplectic, but

$$d\mathbf{q} \wedge d\mathbf{p} = r\,dr \wedge d\theta = d(r^2/2) \wedge d\theta = dI \wedge d\theta,$$

$$I = \frac{1}{2}(\mathbf{q}^2 + \mathbf{p}^2), \qquad \theta = \arctan\frac{\mathbf{p}}{\mathbf{q}},$$

$$\mathbf{q} = \sqrt{2I}\cos\theta, \qquad \mathbf{p} = \sqrt{2I}\sin\theta.$$

Therefore, I, θ are symplectic (or canonical) coordinates called *action-angle* coordinates. The harmonic oscillator $\ddot{\xi} + \omega^2\xi = 0$, if we set $\mathbf{q} = \omega\xi$, $\mathbf{p} = \dot{\xi}$ can be written as a Hamiltonian system with Hamiltonian

$$H = \frac{\omega}{2}(\mathbf{q}^2 + \mathbf{p}^2) = \omega I,$$

and in action-angle coordinates, the equations of motion are

$$\dot{I} = \frac{\partial H}{\partial \theta} = 0, \qquad \dot{\theta} = -\frac{\partial H}{\partial I} = -\omega.$$

So the solutions move in a counterclockwise direction on the circles of constant radius with uniform frequency ω.

Now consider a two degrees of freedom problem with rectangular coordinates $\mathbf{q} = (\mathbf{q}_1, \mathbf{q}_2)$ with their conjugate momentum $\mathbf{p} = (\mathbf{p}_1, \mathbf{p}_2)$. Consider the symplectic extension of polar coordinates in the $\mathbf{q}$ plane — i.e., we wish to change to polar coordinates r, θ in the $\mathbf{q}$-plane. Thus we need to extend this point transformation to a symplectic change of variables. Let R, Θ be conjugate to r, θ. Take as a generating function

$$S = \mathbf{p}_1 r\cos\theta + \mathbf{p}_2 r\sin\theta,$$

so that

$$\mathbf{q}_1 = \frac{\partial S}{\partial \mathbf{p}_1} = r\cos\theta, \qquad \mathbf{q}_2 = \frac{\partial S}{\partial \mathbf{p}_2} = r\sin\theta,$$

$$R = \frac{\partial S}{\partial r} = \mathbf{p}_1\cos\theta + \mathbf{p}_2\sin\theta = \frac{\mathbf{q}_1\mathbf{p}_1 + \mathbf{q}_2\mathbf{p}_2}{r},$$

$$\Theta = \frac{\partial S}{\partial \theta} = -\mathbf{p}_1 r\sin\theta + \mathbf{p}_2 r\cos\theta = \mathbf{q}_1\mathbf{p}_2 - \mathbf{q}_2\mathbf{p}_1.$$

If we think of a particle of mass $m = 1$ moving in the plane, then $\mathbf{p}_1 = \dot{\mathbf{q}}_1$ and $\mathbf{p}_2 = \dot{\mathbf{q}}_2$ are linear momenta in the $\mathbf{q}_1$ and $\mathbf{q}_2$ directions; thus $R = \dot{r}$ is linear momentum in the r direction and $\Theta = \mathbf{q}_1\dot{\mathbf{q}}_2 - \mathbf{q}_2\dot{\mathbf{q}}_1 = r^2\dot{\theta}$ is angular momentum. The inverse transformation is

$$\mathbf{p}_1 = R\cos\theta - \frac{\Theta}{r}\sin\theta, \qquad \mathbf{p}_2 = R\sin\theta + \frac{\Theta}{r}\cos\theta.$$

The Hamiltonian of Kepler's problem in polar coordinates is

$$H = \frac{1}{2}\left(R^2 + \frac{\Theta^2}{r^2}\right) - \frac{\mu}{r}.$$

Since H is independent of θ, it is an ignorable coordinate, and Θ is an integral. These coordinates are used to solve the Kepler problem below.

The Hamiltonian of Kepler's problem in rotating polar coordinates is

$$H = \frac{1}{2}\left(R^2 + \frac{\Theta^2}{r^2}\right) - \Theta - \frac{\mu}{r}.$$

Again H is independent of θ, it is an ignorable coordinate, and Θ is an integral.

Consider the three-body problem in fixed Jacobi coordinates with center of mass at the origin and linear momentum zero. Introduce polar coordinates for $\mathbf{x}_1$ and $\mathbf{x}_2$. That is, let

$$\mathbf{x}_1 = (r_1 \cos\theta_1, r_1 \sin\theta_1), \qquad \mathbf{x}_2 = (r_2 \cos\theta_2, r_2 \sin\theta_2),$$

$$\mathbf{y}_1 = (R_1 \cos\theta_1 - (\Theta_1/r_1)\sin\theta_1, R_1 \sin\theta_1 + (\Theta_1/r_1)\cos\theta_1),$$

$$\mathbf{y}_2 = (R_2 \cos\theta_2 - (\Theta_2/r_2)\sin\theta_2, R_2 \sin\theta_2 + (\Theta_2/r_2)\cos\theta_2),$$

so the Hamiltonian of the three-body problem in Jacobi-polar coordinates becomes

$$\begin{aligned} H = {} & \frac{1}{2M_1}\left\{R_1^2 + \left(\frac{\Theta_1^2}{r_1^2}\right)\right\} + \frac{1}{2M_2}\left\{R_2^2 + \left(\frac{\Theta_2^2}{r_2^2}\right)\right\} - \frac{m_0 m_1}{r_1} \\ & - \frac{m_0 m_2}{\sqrt{r_2^2 + \alpha_1^2 r_1^2 + 2\alpha_1 r_1 r_2 \cos(\theta_2 - \theta_1)}} \\ & - \frac{m_1 m_2}{\sqrt{r_2^2 + \alpha_0^2 r_1^2 - 2\alpha_0 r_1 r_2 \cos(\theta_2 - \theta_1)}} \end{aligned} \tag{3.26}$$

The constants are the same as before. Note that the Hamiltonian depends only on the difference of the polar angles, that is, on $\theta_2 - \theta_1$.

3.7 Solution of the Kepler Problem

Recall that the Hamiltonian of the Kepler problem is

$$H = \frac{\|\mathbf{p}\|^2}{2} - \frac{\mu}{\|\mathbf{q}\|},$$

where $\mathbf{q} \in \mathbb{R}^3$ is the position vector of the particle in a fixed coordinate system, $\mathbf{p} \in \mathbb{R}^3$ is its momentum, and μ is a constant. Also $\mathbf{O} = \mathbf{q} \times \mathbf{p}$, angular momentum, is constant along the solutions, so the three components of $\mathbf{O}$ are integrals. If $\mathbf{O} = 0$, then the motion is collinear. In this case the integrals are elementary and one obtains simple formulas for the solutions.

If $\mathbf{O} \neq 0$, then both $\mathbf{q}$ and $\mathbf{p} = \dot{\mathbf{q}}$ are orthogonal to $\mathbf{O}$, so the motion takes place in the plane orthogonal to $\mathbf{O}$. This plane is called the *invariant plane*. In this case take one coordinate axis, say the last, to point along $\mathbf{O}$, so that the motion is in a coordinate plane. The equations of motion in this coordinate plane have the same form as above, but now $\mathbf{q} \in \mathbb{R}^2$. In the planar problem, only the component of angular momentum perpendicular to the plane is nontrivial; so the problem is reduced to a problem of two degrees of freedom with one integral. Such a problem is solvable "up to quadrature." It turns out that the problem is solvable (well, almost) in terms of elementary functions.

The Hamiltonian of the planar Kepler problem in polar coordinates is

$$H = \frac{1}{2}\left(R^2 + \frac{\Theta^2}{r^2}\right) - \frac{\mu}{r}.$$

Since H is independent of θ, it is an ignorable coordinate, and Θ is an integral. The equations of motion are

$$\dot{r} = R, \qquad \dot{\theta} = \frac{\Theta}{r^2},$$

$$\dot{R} = \frac{\Theta^2}{r^3} - \frac{\mu}{r^2}, \qquad \dot{\Theta} = 0.$$

These equations imply that angular momentum Θ is constant, say c; then

$$\ddot{r} = \dot{R} = c^2/r^3 - \mu/r^2.$$

This is a one degree of freedom equation for r, so it is easily solvable. The equation for r can be solved explicitly.

Assume $c \neq 0$, so the notion is not collinear. Make the changes of variables $u = 1/r$ and $dt = (r^2/c)d\theta$ so that

$$\ddot{r} = \frac{c}{r^2}\frac{d}{d\theta}\left(\frac{c}{r^2}\frac{dr}{d\theta}\right) = c^2u^2\frac{d}{d\theta}\left(u^2\frac{du^{-1}}{d\theta}\right)$$

$$= -c^2u^2u'' - \frac{c^2}{r^3} + \frac{\mu}{r^2} = -c^2u^3 + \mu u^2,$$

or

$$u'' + u = \mu/c^2,$$

where $' = d/d\theta$. This equation is just the nonhomogeneous harmonic oscillator, which has the general solution $u = \mu/c^2(1 + e\cos(\theta - g))$, where e and g are integration constants. Let $f = \theta - g$; then

$$r = \frac{c^2/\mu}{1 + e\cos f}. \tag{3.27}$$

This is the equation of a conic section in polar coordinates. Consider a line ℓ in Figure 3.2 which is perpendicular to the ray at angle g through the origin and at a distance $c^2/e\mu$. Rewrite (3.27) as

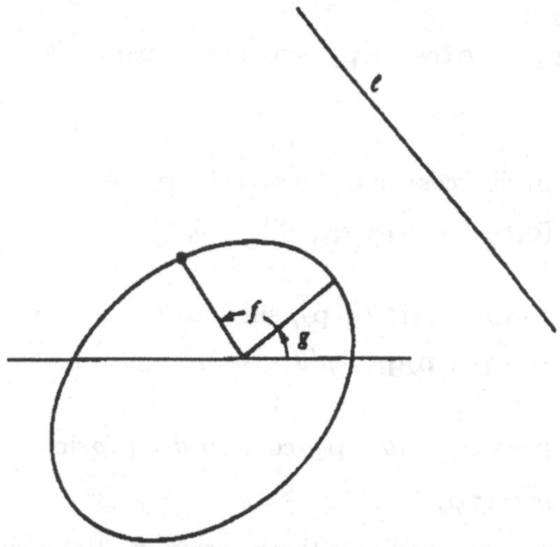

Fig. 3.2. An elliptic orbit

$$r = e\left(\frac{c^2}{e\mu} - r\cos f\right),$$

which says that the distance of the particle to the origin, r, is equal to e times the distance of the particle to the line ℓ , $c^2/e\mu - r\cos f$. This is one of the many definitions of a conic section. One focus is at the origin. The parameter e is the eccentricity. The locus is circle if $e = 0$, an ellipse if $0 < e < 1$, a parabola if $e = 1$, and a hyperbola if $e > 0$.

The point of closest approach, p in Figure 3.2, is called the *perihelion* if the sun is the attractor at the origin or the *perigee* if the earth is. The angle f is called the *true anomaly* and g the *argument of the perihelion (perigee)*.

3.8 Spherical Coordinates

Usually the three dimensional Kepler problem is reduced to the planar problem because conservation of angular momentum implies that the motion takes place in a plane perpendicular to the angular momentum vector. However, sometimes spherical coordinates are useful. This section could be skipped on the first reading.

Consider the standard spherical coordinates (ρ, θ, ϕ), the radius, longitude, and co-latitude, i.e. define spherical coordinates by

$$\mathbf{q}_1 = \rho\sin\phi\cos\theta, \qquad \mathbf{q}_2 = \rho\sin\phi\sin\theta, \qquad \mathbf{q}_3 = \rho\cos\phi.$$

To extend this point transformation use Mathieu generating function

$$S = \mathbf{p}_1 \rho \sin\phi \cos\theta + \mathbf{p}_2 \rho \sin\phi \sin\theta + \mathbf{p}_3 \rho \cos\phi,$$

so

$$\begin{aligned}
P = \frac{\partial S}{\partial \rho} &= \mathbf{p}_1 \sin\phi\cos\theta + \mathbf{p}_2 \sin\phi \sin\theta + \mathbf{p}_3 \cos\phi \\
&= (\mathbf{q}_1\mathbf{p}_1 + \mathbf{q}_2\mathbf{p}_2 + \mathbf{q}_3\mathbf{p}_3)/\rho = \dot{\rho}, \\
\Theta = \frac{\partial S}{\partial \theta} &= -\mathbf{p}_1 \rho \sin\phi \sin\theta + \mathbf{p}_2 \rho \sin\phi\cos\theta \\
&= -\mathbf{p}_1\mathbf{q}_2 + \mathbf{p}_2\mathbf{q}_1 = \rho^2\dot{\theta} \\
\Phi = \frac{\partial S}{\partial \phi} &= \mathbf{p}_1 \rho\cos\phi\cos\theta + \mathbf{p}_2\rho\cos\phi\sin\theta - \mathbf{p}_3\rho\sin\phi \\
&= \rho^2\cos^2\phi\dot{\phi}.
\end{aligned} \tag{3.28}$$

Thus R is radial momentum, and Θ is the $\mathbf{q}_3$-component of angular momentum. From these expressions compute

$$\begin{aligned}
\mathbf{p}_3 &= P\cos\phi - (\Phi/\rho)\sin\phi, \\
P\sin\phi + (\Phi/\rho)\cos\phi &= \mathbf{p}_1\cos\theta + \mathbf{p}_2\sin\theta, \\
\Theta/(\rho\sin\phi) &= -\mathbf{p}_1\sin\theta + \mathbf{p}_2\cos\theta.
\end{aligned}$$

From the last two formulas compute $\mathbf{p}_1^2 + \mathbf{p}_2^2$ without computing $\mathbf{p}_1$ and $\mathbf{p}_2$. You will find that the Hamiltonian of the Kepler problem in spherical coordinates is

$$H = \frac{1}{2}\left\{P^2 + \frac{\Phi^2}{\rho^2} + \frac{\Theta^2}{\rho^2\sin^2\phi}\right\} - \frac{1}{\rho} \tag{3.29}$$

and the equations of motion are

$$\begin{aligned}
\dot{\rho} &= H_P = P, & \dot{P} &= -H_\rho = \frac{\Phi^2}{\rho^3} + \frac{\Theta^2}{\rho^3\sin^2\phi} - \frac{1}{\rho^2}, \\
\dot{\theta} &= H_\Theta = \frac{\Theta}{\rho^2\sin^2\phi}, & \dot{\Theta} &= -H_\theta = 0 \\
\dot{\phi} &= H_\Phi = \frac{\Phi}{\rho^2}, & \dot{\Phi} &= -H_\phi = \left(\frac{\Theta^2}{\rho^2}\right)\frac{\cos\phi}{\sin^3\phi}.
\end{aligned}$$

Clearly, Θ, the $\mathbf{q}_3$-component of angular momentum, is an integral, but so is G defined by

$$G^2 = \left(\frac{\Theta^2}{\sin^2\phi} + \Phi^2\right). \tag{3.30}$$

Later, we shall show that G is the magnitude of total angular momentum.

Now what is the invariant plane in these coordinates? The equation of a plane through the origin is of the form $\alpha \mathbf{q}_1 + \beta \mathbf{q}_2 + \gamma \mathbf{q}_3 = 0$ or in spherical coordinates

$$\alpha \sin\phi\cos\theta + \beta\sin\phi\sin\theta + \gamma\cos\phi = 0$$

or

$$a\sin(\theta - \theta_0) = b\cot\phi.$$

Let the plane meet the $\mathbf{q}_1, \mathbf{q}_2$-plane in a line through the origin with polar angle $\theta = \Omega$ (the *longitude of the node*) and be inclined to the $\mathbf{q}_1, \mathbf{q}_2$-plane by an angle i (the *inclination*).

When $\theta = \Omega$, $\phi = \pi/2$ so we may take $\theta_0 = \Omega$. Let ϕ_m be the minimum ϕ takes on the plane, so $\phi_m + i = \pi/2$. ϕ_m gives the maximum value of $\cot\phi$ and sin has its maximum value of $+1$. Thus $a = b\cot\phi_m$ or $a\sin\phi_m = b\cos\phi_m$. Take $a = \cos\phi_m = \sin i$ and $b = \sin\phi_m = \cos i$. Therefore, the equation of a plane in spherical coordinates with the longitude of the node Ω and inclination i is

$$\sin i \sin(\theta - \Omega) = \cos i \cot\phi.$$

Use (3.30) to solve for Φ and substitute it into the equation for $\dot{\phi}$, then eliminate ρ^2 from the equations for $\dot{\phi}$ and $\dot{\theta}$, to obtain

$$\dot{\phi} = \frac{\Phi}{\rho^2} = \left\{G^2 - \frac{\Theta^2}{\sin^2\phi}\right\}^{1/2} \frac{1}{\rho^2} = \left\{G^2 - \frac{\Theta^2}{\sin^2\phi}\right\}^{1/2} \left\{\frac{\sin^2\phi\dot{\theta}}{\Theta}\right\}.$$

Separate variables and let $\theta = \Omega$ when $\phi = \pi/2$, so that Ω is the longitude of the node. Thus

$$\int_0^\phi \left\{G^2 - \frac{\Theta^2}{\sin^2\phi}\right\}^{-1/2} \sin^{-2}\phi d\phi = \int_\Omega^\theta \Theta^{-1} d\theta = (\theta - \Omega)/\Theta$$

$$-\int_0^u \{G^2 - \Theta^2(1+u^2)\}^{-1/2} du =$$

$$-\Theta^{-1}\int_0^u \{\beta^2 - u^2\}^{-1/2} du =$$

$$\Theta^{-1}\sin^{-1}(u/\beta) =$$

The first substitution is $u = \cot\phi$ and β is defined by $\beta^2 = (G^2 - \Theta^2)/\Theta^2$. Therefore,

$$-\cot\phi = \pm\beta\sin(\theta - \Omega).$$

Finally

$$\cos i \cot \phi = \sin i \sin(\theta - \Omega) \tag{3.31}$$

where

$$\beta^2 = \frac{G^2 - \Theta^2}{\Theta^2} = \tan^2 i = \frac{\sin^2 i}{\cos^2 i}. \tag{3.32}$$

Equation (3.31) is the equation of the invariant plane. The above gives $\Theta = \pm G \cos i$. Since i is the inclination and Θ is the z-component of angular momentum this means that G is the magnitude of total angular momentum. In the above take θ_0 to be Ω the longitude of the node.

3.9 Symplectic Scaling

If instead of satisfying (3.5) transformation $Z = T(t, z)$ satisfies

$$\mu J = \frac{\partial T}{\partial z}^T J \frac{\partial T}{\partial z},$$

where μ is some nonzero constant, then $Z = T(t, z)$ is called a *symplectic transformation (map, change of variables, etc.) with multiplier* μ. Equation (3.3) becomes

$$\dot{Z} = \mu J \nabla_Z H(t, Z) + J \nabla_Z R(t, Z),$$

where all the symbols have the same meaning as before. In the time-independent case, we simply multiply the Hamiltonian by μ. Let us look at a simple application.

Consider a Hamiltonian which has a critical point at the origin, so

$$H(z) = \frac{1}{2} z^T S z + K(z),$$

where S is the Hessian of H at $z = 0$ and K vanishes along with its first and second partial derivatives at the origin. The change of variables $z = \varepsilon Z$ or $Z = T(z) = \varepsilon^{-1} z$ is a symplectic change of variables with multiplier ε^{-2}, so the Hamiltonian becomes

$$H(Z) = \frac{1}{2} Z^T S Z + \varepsilon^{-2} K(\varepsilon Z) = \frac{1}{2} Z^T S Z + O(\varepsilon).$$

In the above, the classical notation $O(\varepsilon)$ of perturbation theory is used. Since K is at least third order at the origin, there is a constant C such that $| \varepsilon^{-2} K(\varepsilon Z) | \le C\varepsilon$ for Z in a neighborhood of the origin and ε small, which is written $\varepsilon^{-2} K(\varepsilon Z) = O(\varepsilon)$. The equations of motion become

$$\dot{Z} = AZ + O(\varepsilon), \qquad A = JS.$$

If $\|Z\|$ is about 1 and ε is small, then z is small: thus the above transformation is useful in studying the equations near the critical point. To the lowest order in ε the equations are linear, so close to the critical point, the linear terms are the most important terms. This is an example of what is called *scaling variables*, and ε is called the *scale parameter*. To avoid the growth of symbols, one often says: scale by $z \to \varepsilon z$, which means replace z by εz everywhere. This would have the effect of changing Z back to z. It must be remembered that scaling is really changing variables.

3.10 Problems

1 Show that a 2×2 matrix T is symplectic if and only if $\det T = 1$.

2 Show that if T and S are symplectic matrices then so are T^{-1}, T^T, and TS.

3 Find the equilibrium points of the Kepler problem in rotating polar coordinates. What do these equilibrium correspond to in non-rotating coordinates? Analyze the linearized equations about this equilibrium point.

4 What is the Hamiltonian of the Kepler problem in rotating-spherical coordinates?

5 The Hamiltonian of the three body problem in Jacobi-polar coordinates (3.26) depends only on $\theta_2 - \theta_1$. Make the symplectic change of variables

$$\phi_1 = \theta_1, \qquad \phi_2 = \theta_2 - \theta_1,$$

$$\Phi_1 = \Theta_1 + \Theta_2, \Phi_2 = \Theta_2.$$

Show that the Hamiltonian is independent of ϕ_1 (it is an ignorable coordinate) and that its conjugate Φ_1, total angular momentum, is a constant.

6 Show that when angular momentum is zero, $\mathbf{O} = 0$, for Kepler's problem that the motion is collinear. Explicitly solve the Kepler's problem in this case.

7 Show that scaling the Hamiltonian by $H \to \nu^{-1}t$ has the effect of scaling time by $t \to \nu t$.

4. Central Configurations

The N-body problem for $N > 2$ has resisted all attempts to be solved; indeed it is generally believed that the problem cannot be integrated in the classical sense. Over the years, many special types of solutions have been found using various mathematical techniques. In this chapter we shall find certain solutions by the time-honored method of guess and test.

These solutions, called central configuration solutions, are important in celestial mechanics for several reasons. As the particles of the N-body problem tend to collision or expand to infinity, the positions of the particles tend to a central configuration — see [70]. For us, the planar central configurations give rise to simple periodic solutions which will be used to establish other periodic solutions in the later chapters.

4.1 Equilibrium Solutions

The simplest type of solution one might look for is the equilibrium solution. From (2.1) or (2.3), an equilibrium solution would have to satisfy

$$\frac{\partial \mathbf{U}}{\partial \mathbf{q}_i} = 0 \qquad for \quad i = 1, \ldots, N. \tag{4.1}$$

However, $\mathbf{U}$ is homogeneous of degree -1, and so by Euler's theorem on homogeneous functions,

$$\sum \mathbf{q}_i \frac{\partial \mathbf{U}}{\partial \mathbf{q}_i} = -\mathbf{U}. \tag{4.2}$$

Since $\mathbf{U}$ is the sum of positive terms, it is positive, but by (4.1) the left-hand side of (4.2) is zero, which is a contradiction. Thus there are no equilibrium solutions of the N-body problem.

4.2 Equations for a Central Configuration

To seek collinear solutions of (2.1), try $\mathbf{q}_i(t) = \phi(t)\mathbf{a}_i$, where the $\mathbf{a}_i$'s are constant vectors in $\mathbb{R}^2$ or $\mathbb{R}^3$ and $\phi(t)$ is a scalar-valued function. Substituting into (2.1) and rearranging, we have

$$|\phi|^3\phi^{-1}\ddot{\phi}m_i\mathbf{a}_i = \sum_{j=1,j\neq i}^{N} \frac{m_im_j(\mathbf{a}_j - \mathbf{a}_i)}{\|\mathbf{a}_j - \mathbf{a}_i\|^3}. \tag{4.3}$$

Since the right-hand side is constant, the left-hand side must be constant also; therefore, (4.3) has a solution if there exist a scalar function $\phi(t)$, a constant λ, and constant vectors $\mathbf{a}_i$ such that

$$\ddot{\phi} = -\frac{\lambda\phi}{|\ \phi\ |^3} \tag{4.4}$$

and

$$-\lambda m_i\mathbf{a}_i = \sum_{j=1,j\neq i}^{N} \frac{m_im_j(\mathbf{a}_j - \mathbf{a}_i)}{\|\mathbf{a}_j - \mathbf{a}_i\|^3} \quad \text{for} \quad i = 1, \ldots, N. \tag{4.5}$$

Equation (4.4) is a simple ordinary differential equation (the one-dimensional Kepler problem!) and so has many solutions. For example, one solution is $\alpha t^{2/3}$, where $\alpha^3 = 9\lambda/2$. This is a solution which goes from zero to infinity as t goes from zero to infinity. The complete analysis of (4.4) is left to the problems. Equation (4.5) is a nontrivial system of nonlinear algebraic equations. The complete solution is known only for $N = 2, 3$, but there are many special solutions known. A configuration of the N particles given by constant vectors $\mathbf{a}_1, \ldots, \mathbf{a}_N$ satisfying (4.5) for some λ is called a *central configuration* (or *c.c.* for short). Central configurations are important in the study of the total collapse of the N-body problem, because it can be shown that the limiting configuration of the N-particles as they tend to a total collapse is a central configuration.

Note that any uniform scaling of a c.c. is also a c.c. In order to measure the size of the N-body system, we define the *moment of inertia* of the system as

$$\mathbf{I} = \frac{1}{2}\sum_{i=1}^{N} m_i\|\mathbf{q}_i\|^2. \tag{4.6}$$

Now (4.5) can be rewritten as

$$\frac{\partial \mathbf{U}}{\partial \mathbf{q}}(\mathbf{a}) + \lambda\frac{\partial \mathbf{I}}{\partial \mathbf{q}}(\mathbf{a}) = 0, \tag{4.7}$$

where $\mathbf{q} = (\mathbf{q}_1, \ldots, \mathbf{q}_N)$ and $\mathbf{a} = (\mathbf{a}_1, \ldots, \mathbf{a}_N)$. The constant λ can be considered as a Lagrange multiplier, so a central configuration is a critical point of the self-potential $\mathbf{U}$ restricted to a constant moment of inertia manifold, $\mathbf{I} = \mathbf{I}_0$, a constant. Fixing $\mathbf{I}_0$ fixes the scale.

Let $\mathbf{a}$ be a central configuration. $\mathbf{U}$ is homogeneous of degree -1 and $\mathbf{I}$ is homogeneous of degree 2. By taking the dot product of the vector $\mathbf{a}$ with the equation in (4.7) and applying Euler's theorem on homogeneous functions, we find that $-\mathbf{U} + 2\lambda\mathbf{I} = 0$ or

$$\lambda = \frac{\mathbf{U(a)}}{2\mathbf{I(a)}} > 0. \tag{4.8}$$

Summing (4.5) on i gives $\sum m_i \mathbf{a}_i = 0$, so the center of mass is at the origin. If A is an orthogonal matrix, either 3×3 in general or 2×2 in the planar case, then clearly $A\mathbf{a} = (A\mathbf{a}_1, \ldots, A\mathbf{a}_N)$ is a c.c. also with the same λ. If $\tau \neq 0$, then $(\tau\mathbf{a}_1, \tau\mathbf{a}_2, \ldots, \tau\mathbf{a}_N)$ is a c.c. also with λ replaced by λ/τ^3. Thus any configuration similar to a c.c. is a c.c. When counting c.c., one only counts similarity classes.

4.3 Relative Equilibrium

Now consider the planar N-body problem, so all the vectors lie in $\mathbb{R}^2$. Identify $\mathbb{R}^2$ with the complex plane $\mathbb{C}$ by considering the $\mathbf{q}_i, \mathbf{p}_i$, etc., as complex numbers. Seek a homographic solution of (2.1) by letting $\mathbf{q}_i(t) = \phi(t)\mathbf{a}_i$, where the $\mathbf{a}_i$'s are constant complex numbers and $\phi(t)$ is a time-dependent complex-valued function. Geometrically, multiplication by a complex number is a rotation followed by a dilation or expansion, i.e., a homography. Thus we seek a solution such that the configuration of the particles is always homographically equivalent to a fixed configuration. Substituting this guess into (2.1) and rearranging gives the same equation (4.3), and the same argument gives equations (4.4), which are now the two-dimensional Kepler problem, and equation (4.5). That is, if we have a solution of (4.5) in which the $\mathbf{a}_i$'s are planar, then there is a solution of the N-body problem of the form $\mathbf{q}_i = \phi(t)\mathbf{a}_i$, where $\phi(t)$ is any solution of the planar Kepler problem, e.g., circular, elliptic, etc.

Consider the N-body problem (3.9) in rotating coordinates $q_1, \ldots, q_N$, $p_1, \ldots, p_N$. An equilibrium solution $q_i = a_i, p_i = b_i$ of the N-body problem in rotating coordinates is called a *relative equilibrium* and must satisfy

$$\frac{b_i}{m_i} + \omega J a_i = 0, \qquad \omega J b_i + \sum_{j=1, j\neq i}^{N} \frac{m_i m_j (a_j - a_i)}{\|a_j - a_i\|^3} \text{ for } i = 1, \ldots, N,$$

and so the a_i's must satisfy

$$-\omega^2 m_i a_i = \sum_{j=1, j\neq i}^{N} \frac{m_i m_j (a_j - a_i)}{\|a_j - a_i\|^3} \text{ for } i = 1, \ldots, N.$$

This is the same as (4.5) with $\lambda = \omega^2$. Thus a planar central configuration gives rise to a relative equilibrium.

The eigenvalues of the linearization of the equations of motion about a relative equilibrium are called the *exponents of the relative equilibrium* and the characteristic polynomial of the linearized equations is called the *characteristic polynomial of the relative equilibrium.* In rotating Jacobi coordinates the four dimensional symplectic subspace where the center of mass is at the origin ($g_N = 0$) and linear momentum is zero ($G_N = 0$) is invariant and (3.23) are the equations of motion on this subspace. Equations (3.23) are linear and the characteristic polynomial of this system is $(\lambda^2 + 1)^2$. Thus, the characteristic polynomial $p(\lambda)$ of a relative equilibrium has $(\lambda^2 + 1)^2$ as a factor.

4.4 Lagrangian Solutions

Consider the c.c. formula (4.5) for the planar three-body problem. Then we seek six unknowns, two components each for $\mathbf{a}_1, \mathbf{a}_2, \mathbf{a}_3$. If we hold the center of mass at the origin, we can eliminate two variables; if we fix the moment of inertia $\mathbf{I}$, we can reduce the dimension by one; and if we identify two configurations which differ by a rotation only, we can reduce the dimension by one again. Thus in theory we can reduce the problem by four dimensions, so that we have a problem of finding critical points of a function on a two-dimensional manifold. This reduction is difficult in general, but there is a trick that works well for the planar three-body problem.

Let $\rho_{ij} = \|\mathbf{q}_i - \mathbf{q}_j\|$ denote the distance between the ith and j^{th} particles. Once the center of mass is fixed at the origin and two rotationally equivalent configurations are identified, then the three variables $\rho_{12}, \rho_{23}, \rho_{31}$ are local coordinates near a non-collinear configuration. The function $\mathbf{U}$ is already written in terms of these variables, since

$$\mathbf{U} = \left(\frac{m_1 m_2}{\rho_{12}} + \frac{m_2 m_3}{\rho_{23}} + \frac{m_3 m_1}{\rho_{31}} \right). \tag{4.9}$$

Let M be the total mass, i.e., $M = \sum m_i$, and assume the center of mass is at the origin; then

$$\begin{aligned}
\textstyle\sum_i \sum_j m_i m_j \rho_{ij}^2 &= \textstyle\sum_i \sum_j m_i m_j \|\mathbf{q}_i - \mathbf{q}_j\|^2 \\
&= \textstyle\sum_i \sum_j m_i m_j \|\mathbf{q}_i\|^2 - 2 \sum_i \sum_j m_i m_j (\mathbf{q}_i, \mathbf{q}_j) \\
&\quad + \textstyle\sum_i \sum_j m_i m_j \|\mathbf{q}_j\|^2 \\
&= 2M\mathbf{I} - 2 \textstyle\sum_i m_i (\mathbf{q}_i, \sum_j m_j \mathbf{q}_j) + 2M\mathbf{I} \\
&= 4M\mathbf{I}.
\end{aligned}$$

Thus if the center of mass is fixed at the origin, then

$$\mathbf{I} = \frac{1}{4M} \sum_i \sum_j m_i m_j \rho_{ij}^2. \tag{4.10}$$

So $\mathbf{I}$ can be written in terms of the mutual distances also. Holding $\mathbf{I}$ fixed is the same as holding $\bar{\mathbf{I}} = \frac{1}{2}(m_1 m_2 \rho_{12}^2 + m_2 m_3 \rho_{23}^2 + m_3 m_1 \rho_{31}^2)$ fixed. Thus the condition for $\mathbf{U}$ to have a critical point on the set $\bar{\mathbf{I}} =$ constant is

$$-\frac{m_i m_j}{\rho_{ij}^2} + \lambda m_i m_j \rho_{ij} = 0 \quad \text{for } (i,j) = (1,2), (2,3), (3,1), \tag{4.11}$$

which clearly has as its only solution $\rho_{12} = \rho_{23} = \rho_{31} = \lambda^{-1/3}$. This solution is an equilateral triangle with λ as a scale parameter; it is attributed to Lagrange.

Theorem 4.4.1. *For any values of the masses, there are two and only two non-collinear central configurations for the three-body problem, namely, the three particles at the vertices of an equilateral triangle. The two solutions correspond to the two orientations of the triangle when labeled by the masses.*

It is trivial to see in these coordinates that the equilateral triangle c.c. is a nondegenerate minimum of the self-potential $\mathbf{U}$.

4.5 Euler-Moulton Solutions

Consider the collinear N-body problem, so $\mathbf{q} = (\mathbf{q}_1, \dots, \mathbf{q}_N) \in \mathbb{R}^N$. Set $S' = \{\mathbf{q} \in \mathbb{R}^N : \mathbf{I}(\mathbf{q}) = 1\}$, an ellipsoid or topological sphere of dimension $N-1$ in $\mathbb{R}^N$; set $G = \{C(\mathbf{q}) = \sum m_i \mathbf{q}_i = 0\}$, a plane of dimension $N-1$ in $\mathbb{R}^N$; and let $S = S' \cap G$, a sphere of dimension $N-2$ in the plane G. (See Figure 4.1, where $N = 3$, S' is a two-sphere, G is a plane, and S is a great circle.) Let $\Delta'_{ij} = \{\mathbf{q} \in \mathbb{R}^N : \mathbf{q}_i = \mathbf{q}_j\}$ and $\Delta' = \bigcup \Delta'_{ij}$; so $\mathbf{U}$ is defined and smooth on $\mathbb{R}^N \backslash \Delta'$. Since Δ' is a union of planes through the origin it intersects S in spheres of dimension $N-3$, denoted by Δ.

Let $\mathcal{U}$ be the restriction of $\mathbf{U}$ to $S \backslash \Delta$; then a critical point of $\mathcal{U}$ is a central configuration. Note that $S \backslash \Delta$ has $N!$ connected components. This is because a component of $S \backslash \Delta$ corresponds to a particular ordering of the $\mathbf{q}_i$'s. That is, to each connected component there is an ordering $\mathbf{q}_{i_1} < \mathbf{q}_{i_2} < \dots < \mathbf{q}_{i_N}$, where $(i_1, i_2, \dots, i_N)$ is a permutation of $1, 2, \dots, N$. There are $N!$ such permutations. Since $\mathcal{U} \to \infty$ as $\mathbf{q} \to \Delta$, the function $\mathcal{U}$ has at least one minimum for each connected component. Thus there are at least $N!$ critical points.

Let a be a critical point of $\mathcal{U}$; then a satisfies (4.5) and $\lambda = \mathbf{U}(a)/2\mathbf{I}(a)$. The derivative of $\mathcal{U}$ at a in the direction $v = (v_1, \dots, v_N) \in T_a S$, where $T_a S$ is the tangent space to S at a, is

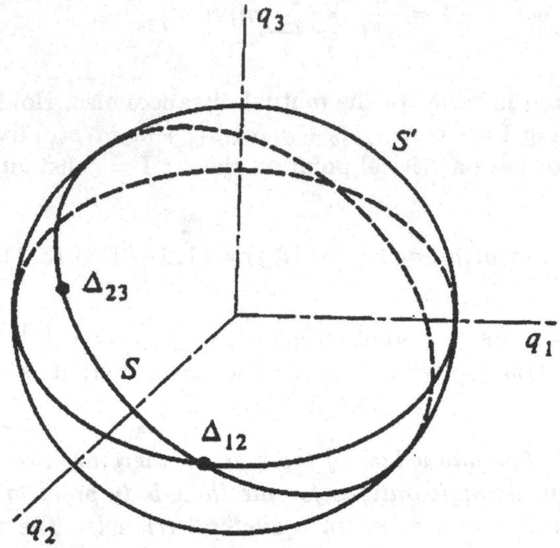

Fig. 4.1. The spaces S' and S for $N = 3$

$$D\mathcal{U}(a)(v) = -\sum \frac{m_i m_j (v_j - v_i)}{| \mathbf{a}_j - \mathbf{a}_i |} + \lambda \sum m_i \mathbf{a}_i v_i, \tag{4.12}$$

and the second derivative is

$$D^2\mathcal{U}(a)(v, w) = 2 \sum \frac{m_i m_j}{| \mathbf{a}_j - \mathbf{a}_i |^3} ((w_j - w_i)(v_j - v_i)) + \lambda \sum m_i w_i v_i. \tag{4.13}$$

From the above, $D^2\mathcal{U}(a)(v, v) > 0$ when $v \neq 0$, so the Hessian is positive definite at a critical point and each such critical point is a minimum of $\mathcal{U}$. Thus there can only be one critical point of $\mathcal{U}$ on each connected component, or there are $N!$ critical points.

In counting the critical points above, we did not remove the symmetry from the problem. The only one-dimensional orthogonal transformation is a reflection in the origin. So, having counted a c.c. and its reflection, we have counted each c.c. twice: thus we have

Theorem 4.5.1. *(Euler-Moulton) There are exactly $N!/2$ collinear central configurations in the N-body problem, one for each ordering of the masses on the line.*

These c.c. are minima of $\mathcal{U}$ only on the line. It can be shown that they are saddle points in the planar problem.

4.6 Central Configuration Coordinates

Consider the planar case. If $(\mathbf{a}_1, \ldots, \mathbf{a}_N)$ is a central configuration, then so is $(\alpha\mathbf{A}\mathbf{a}_1, \ldots, \alpha\mathbf{A}\mathbf{a}_N)$, where α is a nonzero scalar and $\mathbf{A} \in SO(2, \mathbb{R})$ is the set of 2×2 rotation matrices. Since the origin in $\mathbb{R}^{2N}$ is a limit of central configurations, we shall consider it a central configuration also. Thus a central configuration begets the set of central configurations $C_a = \{(\alpha\mathbf{A}\mathbf{a}_1, \ldots, \alpha\mathbf{A}\mathbf{a}_N) : \alpha \in \mathbb{R}, \mathbf{A} \in SO(2, \mathbb{R})\}$. Not all the $\mathbf{a}_j$'s are zero, so assume that $\mathbf{a}_1$ is nonzero. Then $\{\alpha\mathbf{A}\mathbf{a}_1 : \alpha \in \mathbb{R}, \mathbf{A} \in SO(2, \mathbb{R})\}$ is clearly the plane, and this set is isomorphic to C_a. Thus C_a is a two-dimensional linear subspace of $\mathbb{R}^{2N}$.

The equations of motion admit angular momentum,

$$\mathbf{O} = \sum_{j=1}^{N} \mathbf{q}_j^T J \mathbf{p}_j, \tag{4.14}$$

as an integral. Define the critical set $\mathcal{K} \subseteq \mathbb{R}^{4N} \backslash \Delta$ as the set in which $\nabla\mathbf{H}$ and $\nabla\mathbf{O}$ are dependent, i.e.,

$$\mathcal{K} = \{(\mathbf{q}, \mathbf{p}) \in \mathbb{R}^{4N} \backslash \Delta : \alpha\nabla\mathbf{H}(\mathbf{q}, \mathbf{p}) + \beta\nabla\mathbf{O}(\mathbf{q}, \mathbf{p}) = 0, \alpha, \beta \in \mathbb{R}, \alpha^2 + \beta^2 = 1\}. \tag{4.15}$$

Since $\nabla\mathbf{H}$ is never zero and $\nabla\mathbf{O}$ is zero only at the origin, it is enough to look for solutions where both α and β are nonzero. The point $(\mathbf{q}, \mathbf{p}) = (\mathbf{a}, \mathbf{b})$ is in the critical set $\mathcal{K}$ if and only if

$$\alpha\frac{\partial \mathbf{U}}{\partial \mathbf{q}_i} + \beta J \mathbf{p}_j = 0, \qquad \frac{\alpha \mathbf{p}_j}{m_j} = \beta J \mathbf{q}_j. \tag{4.16}$$

If $(\mathbf{a}, \mathbf{b}) \in \mathcal{K}$, then $\mathbf{a} = (\mathbf{a}_1, \ldots, \mathbf{a}_N)$ is a central configuration and $(\mathbf{a}, \mathbf{b})$ is a relative equilibrium.

Let $\mathbf{a} = (\mathbf{a}_1, \ldots, \mathbf{a}_N)$ be a specific central configuration scaled so that $\sum m_j \|\mathbf{a}_j\| = 1$. Let $C' = \{(\alpha\mathbf{A}\mathbf{a}_1, \ldots, \alpha\mathbf{A}\mathbf{a}_N) : \alpha \in \mathbb{R}, \mathbf{A} \in SO(2, \mathbb{R})\}$ as above. Define C' as the subset of $\mathbb{R}^{4N}$ defined by

$$C' = \{(\alpha\mathbf{A}\mathbf{a}_1, \ldots, \alpha\mathbf{A}\mathbf{a}_N; \beta\mathbf{B}m_1\mathbf{a}_1, \ldots, \beta\mathbf{B}m_N\mathbf{a}_N) :$$

$$\alpha, \beta \in \mathbb{R}, \mathbf{A}, \mathbf{B} \in SO(2, \mathbb{R})\}.$$

Proposition 4.6.1. *C' is a four-dimensional invariant linear symplectic subspace of $\mathbb{R}^{4N}$.*

Proof. A symplectic basis for C' is

$$u_1 = (\mathbf{a}_1, \ldots, \mathbf{a}_N; 0, \ldots, 0), \qquad u_2 = (J\mathbf{a}_1, \ldots, J\mathbf{a}_N; 0, \ldots, 0),$$

$$v_1 = (0, \ldots, 0; m_1\mathbf{a}_1, \ldots, m_N\mathbf{a}_N), \; v_2 = (0, \ldots, 0; m_1 J\mathbf{a}_1, \ldots, m_N J\mathbf{a}_N). \tag{4.17}$$

So C' is a four-dimensional linear symplectic subspace of R^{4N}.

For the moment, think of the vectors $\mathbf{q}_j, \mathbf{p}_j$, etc. as complex numbers. Then the set C' is defined by

$$C' = \{(\mathbf{z}\mathbf{a}_1, \dots, \mathbf{z}\mathbf{a}_N; \mathbf{w}m_1\mathbf{a}_1, \dots, \mathbf{w}m_N\mathbf{a}_N) : \mathbf{z}, \mathbf{w} \in \mathbf{C}\}.$$

Let $(\mathbf{z}_0\mathbf{a}_1, \dots, \mathbf{z}_0\mathbf{a}_N; \mathbf{w}_0m_1\mathbf{a}_1, \dots, \mathbf{w}_0m_N\mathbf{a}_N), \mathbf{z}_0, \mathbf{w}_0 \in \mathbf{C}, \mathbf{z}_0 \neq 0$ be any point in C' and let $\mathbf{z}(t), \mathbf{w}(t)$ be the solutions of the Kepler problem

$$\dot{\mathbf{z}} = \mathbf{w}, \qquad \dot{\mathbf{w}} = -\mathbf{z}/ \mid \mathbf{z} \mid^3, \tag{4.18}$$

starting at $\mathbf{z}_0, \mathbf{w}_0$ when $t = 0$. Then it is easy to verify that

$$\mathbf{V}(t) = (\mathbf{q}(t), \mathbf{p}(t)) = (\mathbf{z}(t)\mathbf{a}_1, \dots, \mathbf{z}(t)\mathbf{a}_N; \mathbf{w}(t)m_1\mathbf{a}_1, \dots, \mathbf{w}(t)m_N\mathbf{a}_N)$$

is a solution of the equations of motion of the N-body problem in fixed coordinates and clearly $\mathbf{V}(t) \in C'$ for all t. This proves that C' is invariant.

Proposition 4.6.2. *There exist symplectic coordinates* $(\mathbf{z}, \mathbf{w})$ *for* C' *and symplectic coordinates* $(\mathbf{Z}, \mathbf{W})$ *for* $\mathcal{E}' = \{x \in \mathbb{R}^{4N} : \{x, C'\} = 0\}$ *so that* $(\mathbf{z}, \mathbf{Z}; \mathbf{w}, \mathbf{W})$ *are symplectic coordinates for* $\mathbb{R}^{4N}$ *and the Hamiltonian of the* N*-body problem,* $H(\mathbf{z}, \mathbf{Z}; \mathbf{w}, \mathbf{W})$*, has the properties*

$$\begin{aligned} \partial \mathbf{H}(\mathbf{z}, 0; \mathbf{w}, 0)/\partial \mathbf{Z} &= 0, \\ \partial \mathbf{H}(\mathbf{z}, 0; \mathbf{w}, 0)/\partial \mathbf{W} &= 0, \\ \mathbf{H}(\mathbf{z}, 0; \mathbf{w}, 0) = \mathbf{H}_K(\mathbf{z}, \mathbf{w}) &= \frac{1}{2}\|\mathbf{w}\|^2 - \frac{1}{\|\mathbf{z}\|}. \end{aligned} \tag{4.19}$$

Proof. Since C' is symplectic, $\mathcal{E}'$ is also symplectic and $\mathbb{R}^{4N} = C' \oplus \mathcal{E}'$ — see Meyer and Hall [51], page 43. Thus the vectors in (4.17) can be extended to a symplectic basis $u_1, \dots, u_{2N}; v_1, \dots, v_{2N}$, with $u_3, \dots, u_{2N}; v_3, \dots, v_{2N}$ a symplectic basis for $\mathcal{E}'$. Given this basis, let $\mathbf{z}, \mathbf{w}$ be symplectic coordinates for C' as in the proof of Proposition 4.6.1 and $\mathbf{Z}, \mathbf{W}$ be symplectic coordinates for $\mathcal{E}'$. The first two equations in (4.19) simply say that C' is invariant, and the last says that in the $\mathbf{z}, \mathbf{w}$ coordinates in C', the motion is that of the Kepler problem (4.18). $\mathbf{H}_K$ is just the Hamiltonian of the Kepler problem.

Remark: The above discussion heavily used complex multiplication and thus is valid for the planar problem only. One can use real numbers for the general case of $\mathbb{R}^n$. In this case the corresponding C' would be a two-dimensional invariant linear symplectic subspace of $\mathbb{R}^{2nN}$. The coordinates $\mathbf{z}$ and $\mathbf{w}$ in (4.19) would be one-dimensional.

For $n = 4$ (respectively, $n = 8$), one can use Hamilton's quaternions (Cayley numbers). In these cases the corresponding C' would be an eight- (a sixteen-) dimensional, invariant linear symplectic subspace of $\mathbb{R}^{8N}$ ($\mathbb{R}^{16N}$). The coordinates $\mathbf{z}$ and $\mathbf{w}$ in (4.19) would be four- (eight-) dimensional.

The argument given above uses only the homogeneity of the force field and so works for inverse power laws in general.

Now consider the problem in rotating coordinates. Let $a = (a_1, \ldots, a_N)$ be a central configuration and $C = \{(\alpha A a_1, \ldots, \alpha A a_N) : \alpha \in \mathbb{R}, A \in SO(2,\mathbb{R})\}$ as above. Define $\mathcal{C}$ as the subset of $\mathbb{R}^{4N}$ defined by

$$\mathcal{C} = \{(\alpha A a_1, \ldots, \alpha A a_N; \beta B m_1 a_1, \ldots, \beta B m_N a_N) :$$

$$\alpha, \beta \in \mathbb{R}, A, B \in SO(2,\mathbb{R})\}.$$

The same reasoning yields the following Proposition.

Proposition 4.6.3. *$\mathcal{C}$ is a four-dimensional invariant linear symplectic subspace of $\mathbb{R}^{4N}$. Moreover, there exist symplectic coordinates (z,w) for $\mathcal{C}$ and symplectic coordinates (Z,W) for $\mathcal{E} = \{x \in \mathbb{R}^{4N} : \{x,\mathcal{C}\} = 0\}$ so that $(z,Z;w,W)$ are symplectic coordinates for $\mathbb{R}^{4N}$ and the Hamiltonian of the N-body problem, $H(z,Z;w,W)$, has the properties*

$$\frac{\partial H(z,0;w,0)}{\partial Z} = 0, \qquad \frac{\partial H(z,0;w,0)}{\partial W} = 0,$$
$$H(z,0;w,0) = H_K(z,w) = \frac{1}{2}\|w\|^2 - z^T K w - \frac{1}{\|z\|}. \tag{4.20}$$

Now consider angular momentum, first in fixed coordinates and then in rotating coordinates.

Proposition 4.6.4. *In the symplectic coordinates $(\mathbf{z},\mathbf{Z};\mathbf{w},\mathbf{W})$ of Proposition 4.6.2, angular momentum has the form*

$$\mathbf{O} = \mathbf{z}^T J \mathbf{w} + (\mathbf{Z}^T, \mathbf{W}^T)\mathbf{B}(\mathbf{Z}^T, \mathbf{W}^T)^T, \tag{4.21}$$

where $\mathbf{B}$ is a skew-symmetric matrix of dimension $(4N-4)\times(4N-4)$. In the symplectic coordinates $(z,Z;w,W)$ of Proposition 4.6.3, angular momentum has the form

$$O = z^T J w + (Z^T, W^T) B (Z^T, W^T)^T, \tag{4.22}$$

where B is a skew-symmetric matrix of dimension $(4N-4)\times(4N-4)$.

Proof. Use complex notation again and consider the case of rotating coordinates, since both cases are essentially the same. Recall that $O = \sum q_j^T J p_j = \sum \Im \bar{q}_j p_j$. When $Z = W = 0$, we have $q_j = z a_j$ and $p_j = m_j w a_j$, so $O = \Im \bar{z} w \sum m_j \mid a_j \mid^2 = \Im \bar{z} w$. (Here $\Im$ stands for the imaginary part of a complex number.)

Since the Hamiltonian on the invariant subspaces C' (respectively, $\mathcal{C}$) is just the Hamiltonian of the Kepler problem in fixed coordinates (rotating coordinates), there are lots of special coordinates which simplify the equations of motion. One useful special coordinate system is the Poincaré elements — see Meyer and Hall [51] or Szebehely [86]. These coordinates are valid in a neighborhood of the circular orbits. In a non-rotating frame, the Hamiltonian of the Kepler problem in Poincaré elements is

$$\mathbf{H}_K(\mathbf{Q}_1, \mathbf{Q}_2, \mathbf{P}_1, \mathbf{P}_2) = -1/2\mathbf{P}_1^2, \tag{4.23}$$

and in the rotating frame it is

$$H_K(Q_1, Q_2; P_1, P_2) = -1/2P_1^2 - P_1 + \frac{1}{2}(Q_2^2 + P_2^2). \tag{4.24}$$

In the above, $\mathbf{Q}_1$ is an angular variable defined modulo 2π and the other variables are rectangular variables. Angular momentum in these variables is given by

$$\mathbf{O}_K = \mathbf{P}_1 - (\mathbf{Q}_2^2 + \mathbf{P}_2^2), \qquad O_K = P_1 - (Q_2^2 + P_2^2). \tag{4.25}$$

Note that $\mathbf{Q}_2 = \mathbf{P}_2 = 0$ and $Q_2 = P_2 = 0$ correspond to the circular orbits of the Kepler problem.

Proposition 4.6.5. *Consider a fixed central configuration* **a** *of the N-body problem in fixed coordinates. This central configuration gives rise to a periodic solution $\mathbf{V}_a(t)$ where the bodies uniformly rotate about the center of mass on circular orbits. Let the period of $\mathbf{V}_a(t)$ be $\mathbf{T}_a$ and the energy of $\mathbf{V}$ be $\mathbf{H}_a$. In the fixed energy surface $\mathbf{H} = \mathbf{H}_a$, there is an invariant three-dimensional manifold containing $\mathbf{V}_a$ and filled with periodic solutions, all of period exactly $\mathbf{T}_a$. In fact, this set is the subset of C' with $\mathbf{H} = \mathbf{H}_K$ fixed at $\mathbf{H}_a$.*

Proof. It is well known that the period of the elliptic solutions depends only on the value of the energy. One can also integrate the problem in Poincaré elements.

We have shown that $(\lambda^2+1)^2$ is a factor of the characteristic polynomial of a relative equilibrium, but there are other known factors. Consider the system on the invariant subspace given in Proposition 4.6.3. The Hamiltonian of the Kepler problem in a rotating frame in polar coordinates is

$$H_K = \frac{1}{2}\left(R^2 + \frac{\Theta^2}{r^2}\right) - \Theta - \frac{1}{r}. \tag{4.26}$$

The circular orbit when angular momentum $\Theta = +1$ is an equilibrium point given by θ arbitrary, $r = 1, \Theta = 1, R = 0$. The linearized equations about this periodic solution are

$$\dot{\theta} = \Theta - 2r, \qquad \dot{\Theta} = 0, \qquad \dot{r} = R, \qquad \dot{R} = -r + 2\Theta. \tag{4.27}$$

The characteristic equation of this system is $\lambda^2(\lambda^2+1)$. Thus we have shown

Proposition 4.6.6. *The characteristic polynomial $p(\lambda)$ of a relative equilibrium has the factor $\lambda^2(\lambda^2+1)^3$.*

Let $p(\lambda) = \lambda^2(\lambda^2+1)^3 r(\lambda)$. If $r(\lambda)$ does not have zero as a root, then the relative equilibrium will be called *nondegenerate*, and if $r(\lambda)$ does not have a zero of the form ni, where n is an integer, then the relative equilibrium is called *nonresonant*. For the two-body problem we have $p(\lambda) = \lambda^2(\lambda^2+1)^3$, so the relative equilibria are nonresonant. Moulton's collinear relative equilibria are nondegenerate for all N — see Pacella [62]. The Lagrange equilateral triangle relative equilibria and the Euler collinear relative equilibria for the three-body problem are nonresonant by the analysis in Siegel and Moser [81], Section 18.

4.7 Problems

1 When $\phi(t)$ is a real scalar, equation (4.4) is a one degree of freedom Hamiltonian systems. It is the one dimensional Kepler problem with Hamiltonian $\mathbf{H} = \frac{1}{2}\|\mathbf{p}\|^2 - 1/\|\mathbf{q}\|$. Plot the level curves of $\mathbf{H}$ in the $\mathbf{q}, \mathbf{p}$-plane. In the equation $\mathbf{H} = h$, solve for $\mathbf{p} = \dot{\mathbf{q}}$ and separate variables in this first order equation for $\mathbf{q}$. Thus you have solved the one dimensional Kepler problem.

2 Show that for any values of the masses, there are two and only two noncoplanar c.c. for the four-body problem, namely, the regular tetrahedron configuration with two orientations. (Hint: Read the section on Lagrangian solutions again.)

3 Consider the spatial Kepler problem in rotating coordinates (rotating about the $\mathbf{q}_3$ axis). Find the relative equilibrium and its characteristic polynomial.

4 State and prove the spatial generalization of Proposition 4.6.6. (Hint: Use spherical coordinates.)

5. Symmetries, Integrals, and Reduction

Many of the mathematical models of physical systems have special symmetry properties which are introduced to simplify the analysis. The N-body problem has many symmetries due to the facts that the particles are assumed to be point masses and that Newton's law of gravity assumes that space is homogeneous and isotropic. Symmetries often introduce degenerates into the problem, which can cause problems with the analysis unless the symmetries are exploited correctly. This chapter is devoted to understanding the main symmetry of the N-body problem, namely, its invariance under the group of Euclidean motions, and to exploiting this symmetry. We will touch on some topics from differential topology but not belabor the subject. Some proofs will be outlined or given by reference.

As stated above, the N-body problem has many symmetries. In general, there are two types of symmetries which occur in differential equations, namely, discrete and continuous symmetries.

First consider some examples of systems with discrete symmetries. The Kepler problem is a simple example. Recall that the equations of motion are

$$\dot{\mathbf{q}} = \mathbf{H_p} = \mathbf{p}, \qquad \dot{\mathbf{p}} = -\mathbf{H_q} = -\frac{\mu\mathbf{q}}{\|\mathbf{q}\|^3},$$

where $\mathbf{H}$ is the Hamiltonian

$$\mathbf{H} = \frac{1}{2}\|\mathbf{p}\|^2 - \frac{\mu}{\|\mathbf{q}\|}.$$

The Hamiltonian is invariant under the reflection $(\mathbf{q},\mathbf{p}) \longrightarrow (\mathbf{q},-\mathbf{p})$, which implies that if $(\mathbf{q}(t),\mathbf{p}(t))$ is a solution, then so is $(\mathbf{q}(-t),-\mathbf{p}(-t))$. This is a property shared by many Hamiltonian systems, since many are quadratic in the momentum (e.g., the N-body problem). Such systems are called *reversible*. Notice that this transformation is *anti-symplectic*, i.e., it takes the symplectic form $d\mathbf{q} \wedge d\mathbf{p}$ into its negative, or satisfies $A^T JA = -J$, where $A = diag\{I,-I\}$. The Hamiltonian is also invariant under the transformation $(\mathbf{q},\mathbf{p}) \longrightarrow (-\mathbf{q},\mathbf{p})$, with similar implications.

As a second example, consider the restricted three-body problem

$$H = \frac{\|p\|^2}{2} - q^T Jp - U,$$

where $q, p \in \mathbb{R}^2$ are conjugate and U is the self-potential

$$U = \frac{\mu}{((q_1 - 1 + \mu)^2 + q_2^2)^{1/2}} + \frac{1-\mu}{((q_1 + \mu)^2 + q_2^2)^{1/2}}.$$

It is invariant under the anti-symplectic transformation $(q_1, q_2, p_1, p_2) \longrightarrow (q_1, -q_2, -p_1, p_2)$. This implies that if $(q_1(t), q_2(t), p_1(t), p_2(t))$ is a solution, then so is $(q_1(-t), -q_2(-t), -p_1(-t), p_2(-t))$. This in turn implies that if a solution of the restricted three-body problem crosses the line of syzygy (the line joining the primaries) perpendicularly at times $t = 0$ and $t = T \neq 0$, then the solution is $2T$-periodic. This criterion has been used many times to find periodic solutions of the restricted three-body problem and other similar problems. Both of these examples are called discrete symmetries.

This book will *not* be concerned with establishing periodic solutions by exploiting discrete symmetries. First, the practical reason is that establishing periodic solutions using this kind of symmetry is the subject of a book of equal or greater length! Second, the theoretical reasons are that not all systems admit this type of symmetry and quite often this type of argument gives no stability information.

Now consider some systems which admit continuous symmetries. The Kepler problem is invariant under rotations as well as under reflection, that is, if A is a rotation matrix, then the Hamiltonian of the Kepler problem in invariant under the transformation $(\mathbf{q}, \mathbf{p}) \longrightarrow (A\mathbf{q}, A\mathbf{p})$. Note that this transformation is symplectic. This invariance implies that if $(\mathbf{q}(t), \mathbf{p}(t))$ is a solution of the Kepler problem, then so is $(A\mathbf{q}(t), A\mathbf{p}(t))$. This type of symmetry is called a continuous symmetry because there is a continuum of rotations.

The Hamiltonian of the N-body problem,

$$\mathbf{H} = \sum_{i=1}^{N} \frac{\|\mathbf{p}_i\|^2}{2m_i} - \sum_{1 \le i < j \le N} \frac{m_i m_j}{\|\mathbf{q}_i - \mathbf{q}_j\|},$$

is invariant under Euclidean motions, i.e., rotations and translations. The Hamiltonian of the N-body problem is invariant under the transformation $(\mathbf{q}_1, \ldots, \mathbf{q}_N, \mathbf{p}_1, \ldots, \mathbf{p}_N) \longrightarrow (A\mathbf{q}_1 + b, \ldots, A\mathbf{q}_N + b, A\mathbf{p}_1, \ldots, A\mathbf{p}_N)$, where A is a rotation matrix and b is a vector. This implies that a Euclidean displacement of a solution is a solution also. This seems reasonable, for if the universe consisted of just N point masses, then there would not be any preferred coordinate frame, and one frame is as good as the next.

Continuous symmetries of Hamiltonian systems imply the existence of integrals. Symmetries and integrals cause degeneracies in perturbation analysis. Therefore, we must study the implications of the symmetries and integrals on the dynamics.

5.1 Group Actions and Symmetries

We will need some basic terminology of differentiable manifolds and Lie groups. The examples in this book are very concrete, but the general terminology is important for keeping the concepts clear. Here we shall give loose definitions, and the reader is referred to a modern book on differentiable manifolds, such as [1, 35], for complete details.

A *manifold (of dimension n)* is a topological space $\mathfrak{M}$ such that each point of $\mathfrak{M}$ has a neighborhood which is homeomorphic to an open set in $\mathbb{R}^n$. That is, if people living in this space were sufficiently myopic, they would think they were living in $\mathbb{R}^n$. This explains why many hidebound people thought the earth was flat. Standard manifolds are: $\mathbb{R}^n$, Euclidean space; $\mathbb{S}^n = \{x \in \mathbb{R}^{n+1} : \|x\| = 1\}$, the n-sphere; $\mathbb{T}^n = \mathbb{S}^1 \times \cdots \times \mathbb{S}^1$, the n-torus; the Möbius strip; any smooth surface without self-intersection. If $\zeta \in \mathfrak{M}$ is such point, then there are a neighborhood W of ζ and a homeomorphism $\phi : W \longrightarrow V \subset \mathbb{R}^n$. The pair (W, ϕ) is called a *chart at* ζ or simply a *local coordinate system at* ζ. Think of the surface of the earth as the two-sphere, $\mathbb{S}^2$. Then a good atlas has enough charts (maps) so that every point on the earth is in at least one chart. Since the charts are on a flat page, they can be considered in $\mathbb{R}^2$. The chart is a picture of a homeomorphism of a part of the surface of the earth into $\mathbb{R}^2$.

A *differentiable manifold* is simply a manifold with a collection of local coordinate systems such that every change of coordinates is differentiable with a differentiable inverse. This collection is called an *atlas*. A differential manifold is the natural place to do differential analysis. Functions, differential equations, differential forms, etc. are *differentiable* or *smooth* if they are differentiable in all coordinate systems of the atlas.

A *Lie group* is a group $\mathfrak{G}$ whose underlying set is a manifold such that multiplication and taking inverses are smooth operations — see [35]. There is no loss of generality in thinking of a Lie group as a closed subgroup of the group of all invertible $n \times n$ matrices with matrix multiplication as the group product.

Some of the standard examples of Lie groups are:

- $GL_n = GL(n, \mathbb{R})$: The group of all real $n \times n$ invertible matrices. It is called the *general linear group*.
- $O_n = O(n, \mathbb{R})$: The group of all real $n \times n$ orthogonal matrices. It is called the *orthogonal group*.
- $SL_n = SL(n, \mathbb{R})$: The group of all real $n \times n$ matrices with determinant equal to $+1$. It is called the *special linear group*.
- $SO_n = SO(n, \mathbb{R}) = O_n \cap SL_n$: It is called the *special orthogonal group*.
- $Sp_n = Sp(n, \mathbb{R})$: The group of all real $2n \times 2n$ symplectic matrices. It is called the *symplectic group*.
- $U_n = U(n, \mathbb{C})$: The group of all complex $n \times n$ unitary matrices. It is called the *unitary group*.

- $\mathbb{R}^n$: with the group product being vector addition.

A *(smooth) action of a Lie group* $\mathfrak{G}$ *on manifold* $\mathfrak{M}$ is a map $\Psi : \mathfrak{G} \times \mathfrak{M} \longrightarrow \mathfrak{M}$ such that

- For each $\gamma \in \mathfrak{G}$, the map $\Psi(\gamma, \cdot) : \mathfrak{M} \longrightarrow \mathfrak{M}$ is a diffeomorphism.
- The map $\Psi(e, \cdot) : \mathfrak{M} \longrightarrow \mathfrak{M}$ is the identity map, where e is the identity element in $\mathfrak{G}$.
- $\Psi(\gamma_1, \Psi(\gamma_2, \zeta)) = \Psi(\gamma_1 \cdot \gamma_2, \zeta)$ for all $\gamma_1, \gamma_2 \in \mathfrak{G}$ and all $\zeta \in \mathfrak{M}$.

When there is only one action under discussion, we use the more compact notation $\Psi(\gamma, \zeta) = \gamma\zeta$, so last property in the above can be written $\gamma_1(\gamma_2\zeta) = (\gamma_1\gamma_2)\zeta$.

Let $\mathfrak{G}$ be any of the matrix groups listed above. Then $\Psi(\gamma, \zeta) = \gamma\zeta$ is an action on $\mathbb{R}^n$, where $\gamma\zeta$ is just the usual product of the matrix γ and the vector ζ. An action of the additive group $\mathbb{R}^n$ on $\mathbb{R}^n$ is $\Psi(\gamma, \zeta) = \gamma + \zeta$.

When confused, think that the manifold is Euclidean space $\mathbb{R}^n$, *the group is rotations* SO_n, *and the action is matrix multiplication!*

Consider an ordinary differential equation on $\mathfrak{M}$ of the form

$$\dot{z} = f(z), \tag{5.1}$$

where z is a local coordinate system on $\mathfrak{M}$ and f is smooth in the coordinate system. An autonomous differential equation on $\mathfrak{M}$ is thought of as a smooth vector field on $\mathfrak{M}$. At any point, the differential equation defines a solution and the derivative of that solution is thought of as a tangent vector at that point.

This tangent vector at z is, of course, $f(z)$. For us, smooth vector fields and autonomous differential equations are the same. Let $\phi(t, \zeta)$ be the solution of (5.1) satisfying $\phi(0, \zeta) = \zeta$. We will call $\phi(t, \zeta)$ the *general solution* and sometimes write $\phi_t = \phi(t, \cdot)$. We assume for simplicity of the discussion that $\phi(t, \zeta)$ is defined for all $t \in \mathbb{R}, \zeta \in \mathfrak{M}$. In this case, we say that ϕ defines a *flow on* $\mathfrak{M}$, that is, ϕ satisfies the following:

- For each $t \in \mathbb{R}$, the map $\phi_t : \mathfrak{M} \longrightarrow \mathfrak{M}$ is a diffeomorphism.
- The map $\phi_0 : \mathfrak{M} \longrightarrow \mathfrak{M}$ is the identity map.
- $\phi_t \circ \phi_\tau \equiv \phi_{t+\tau}$, i.e., $\phi(t, \phi(\tau, \zeta)) = \phi(t + \tau, \zeta)$ for all $t, \tau \in \mathbb{R}$ and all $\zeta \in \mathfrak{M}$.

Note that a flow is an action of the additive group $\mathbb{R}$ on $\mathfrak{M}$ and so we sometimes write $\phi(t, \zeta) = t\zeta$. An autonomous differential equation like (5.1) defines a flow and conversely. Given $\phi(t, \zeta)$, the differential equation defining it is (5.1) with $f(z) = (\partial\phi(t, z)/\partial t)\mid_{t=0}$.

A *symmetry* of (5.1) is a diffeomorphism $g : \mathfrak{M} \longrightarrow \mathfrak{M}$ such that if $\phi(t)$ is a solution of (5.1), then so is $g(\phi(t))$, i.e., g takes solutions to solutions. That is, g is a symmetry if

$$\frac{dg(\phi(t))}{dt} = f(g(\phi(t))).$$

By differentiating this last expression and then setting $t = 0$, we prove that g is a symmetry if and only if

$$\frac{\partial g}{\partial z}(z)f(z) = f(g(z)) \qquad (Dgf = f \circ g).$$

A group $\mathfrak{G}$ (actually the action Ψ of $\mathfrak{G}$) is a *symmetry group* of the equation (5.1) if for each $g \in \mathfrak{G}$, the diffeomorphism $\Psi(g, \cdot) : \mathfrak{M} \longrightarrow \mathfrak{M}$ is a symmetry of (5.1).

Example 1: Consider the Aristotelian central force problem

$$\dot{z} = -\frac{z}{\|z\|^3}, \qquad z \in \mathbb{R}^3 \setminus \{0\}.$$

(In Aristotle's physics, force was proportional to velocity.) This problem is clearly rotationally symmetric. In the fancy words given above, let Ψ be the action of the rotation group SO_3 on $\mathbb{R}^3 \setminus \{0\}$ by matrix multiplication and let f be the right hand side of the equation above, $f(z) = -z/\|z\|^3$. Then for each $A \in SO_3, z \in \mathbb{R}^3 \setminus \{0\}$, we have

$$f(\Psi(A,z)) = f(Az) = \frac{-Az}{\|Az\|^3} = A\left(\frac{-z}{\|z\|^3}\right) = Af(z) = \frac{\partial \Psi(A,z)}{\partial z} f(z).$$

Note that for the usual matrix action, $f(Az) = Af(z)$ suffices.

Example 2: Consider the system in the plane given by

$$\dot{z} = Jz + z(1 - \|z\|^2), \qquad z \in \mathbb{R}^2.$$

One can verify directly that this system is SO_2-invariant as was done above, or change the system to polar coordinates

$$\dot{r} = r(1 - r^2), \qquad \dot{\theta} = -1, \tag{5.2}$$

and note that these equations do not depend on θ. The phase portrait is given in Figure 5.1, where only four orbits are shown — the origin, the limit cycle, and two orbits asymptotic to the limit cycle. All the other orbits are obtained by rotating the asymptotic orbits.

Proposition 5.1.1. *Let $\mathfrak{G}$ be a symmetry group with action Ψ of the differential equation (5.1) and let ϕ be the general solution of (5.1). Then*

$$\Psi(\gamma, \phi(t,\zeta)) = \phi(t, \Psi(\gamma,\zeta)) \quad (or\ \gamma(t\zeta) = t(\gamma\zeta))$$

for all $\gamma \in \mathfrak{G}, t \in \mathbb{R}, \zeta \in \mathfrak{M}$.

This simply means that the map $\Psi_\gamma : \zeta \longrightarrow \gamma\zeta$ takes a solution to a solution. In the Aristotelian central force problem, if we rotate a solution, we still have a solution.

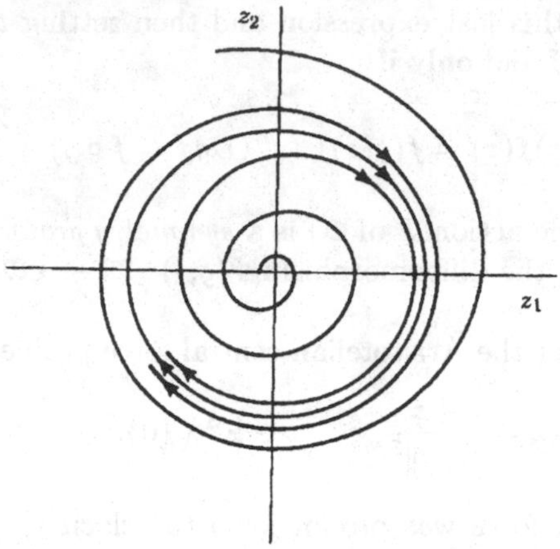

Fig. 5.1. Phase portrait of second example

One of the ways to handle a system which has a symmetry group is to study the system on the quotient space obtained by identifying symmetric points. This quotient space is usually called the *reduced space*. The advantage is that the reduced space is of lower dimension and is sometimes considerably simpler. The disadvantages are, first, that a quotient space may not be a nice space (not a manifold) and, second, that it can be quite difficult to write the equations on the reduced space. It is important for us to understand the reduced space, since the periodic solutions of the N-body problem established in the later chapters live on a reduced space.

Let Ψ be an action of $\mathfrak{G}$ on $\mathfrak{M}$. Call two points $\zeta_1, \zeta_2 \in \mathfrak{M}$ equivalent and write $\zeta_1 \sim \zeta_2$ if there is a $\gamma \in \mathfrak{G}$ such that $\gamma\zeta_1 = \zeta_2$ (*i.e.*, $\Psi(\gamma, \zeta_1) = \zeta_2$). Let $[\zeta] = \{w \in \mathfrak{M} : \zeta \sim w\}$, so $[\zeta]$ is the set of all points equivalent to ζ, the *equivalence class of* ζ. The reduced space is the set of all equivalence classes, i.e., $\mathfrak{M}_R = \mathfrak{M}/\mathfrak{G} = \{[\zeta] : \zeta \in \mathfrak{M}\}$. One places the quotient topology on $\mathfrak{M}_R$ so one has a concept of closeness — $[\zeta]$ and $[w]$ are close if there are $\zeta' \in [\zeta]$ and $w' \in [w]$ with ζ' and w' close. However, this quotient space may not be a manifold in general. That is the bad news; the good news is that the flow ϕ naturally defines a topological flow ϕ_R on the quotient space by

$$\phi_R(t, [\zeta]) = [\phi(t, \zeta)].$$

In words, the flow ϕ takes an equivalence class into an equivalence class. This follows from the definitions and from Proposition 5.1.1. Define the projection map by $\Pi : \mathfrak{M} \longrightarrow \mathfrak{M}_R : \zeta \longrightarrow [\zeta]$. The action of projecting by Π will sometimes be loosely referred to as "dropping down."

Example 1 (continued): The symmetry group for the Aristotelian central force problem is SO_3 acting on $\mathbb{R}^3 \setminus \{0\}$. Given any point in $\zeta \in \mathbb{R}^3 \setminus \{0\}$ at distance ρ from the origin, then ζ can be rotated to the point $(\rho, 0, 0)$ on the positive ray of the first coordinate axis. Thus the reduced space is an open ray, which is a manifold. If we put the Aristotelian central force problem in spherical coordinates, we note that the equations do not depend on the two angles, and the radius equation becomes $\dot{\rho} = -1/\rho^2$, which defines the flow on the reduced space. In this flow, all the orbits tend to the origin. (Aren't you glad Aristotle was wrong?)

Example 2 (continued): In the second example above, the reduced space is a closed ray (a manifold with boundary) and the flow is given by the equation $\dot{r} = r(1 - r^2)$. The flow has fixed points at 0 and 1, and all the other orbits are asymptotic to 1.

In the second example, the reduced space fails to be a manifold because the origin is a boundary point. Far worse things can happen, so we need a criterion that ensures a nice reduced space.

An action $\Psi : \mathfrak{G} \times \mathfrak{M} \longrightarrow \mathfrak{M}$ is *free* if for each $\zeta \in \mathfrak{M}$, the only solution of $\Psi(\gamma, \zeta) = \gamma\zeta = \zeta$ is $\gamma = e$, the identity element of $\mathfrak{G}$. Note that neither of our examples is free. In the first example, for each $z \in \mathbb{R}^3 \setminus \{0\}$ there is a nontrivial rotation that has z as its axis and so this rotation leaves z fixed. In the second example, all elements of the group leave $\zeta = 0$ fixed, so the second action is not free.

The action is *proper* if the map $\tilde{\Psi} : \mathfrak{G} \times \mathfrak{M} \longrightarrow \mathfrak{M} \times \mathfrak{M} : (\gamma, \zeta) \longrightarrow (\zeta, \Psi(\gamma, \zeta))$ is a proper map, i.e., the inverse image of a compact set is compact. Fortunately, if the group $\mathfrak{G}$ is compact, the action is proper, and our main example, SO_n, is compact.

Proposition 5.1.2. *If the action $\Psi : \mathfrak{G} \times \mathfrak{M} \longrightarrow \mathfrak{M}$ is free and proper, then the reduced space is a manifold.*

This is not a very sharp result, since the first example is not free, but the reduced space is a nice manifold. The *isotropy group* of $\zeta \in \mathfrak{M}$ is $\mathfrak{G}_\zeta = \{\gamma \in \mathfrak{G} : \gamma\zeta = \zeta\}$. In the first example, the isotropy group of any $z \in \mathbb{R}^3 \setminus 0$ is the group of rotations about z, which is essentially SO_2. In general, if the isotropy group depends smoothly on the point, the quotient space is still a manifold. See [35] for details and proofs.

By Proposition 5.1.1, an orbit in the full space $\mathfrak{M}$ is projected onto an orbit on the reduced space $\mathfrak{M}_R$, and so invariant sets are projected onto invariant sets. Sometimes the projected set is simpler than the original. In Example 2 above, the limit cycle $r = 1, \theta$ arbitrary is projected onto the critical point $r = 1$. An invariant set for a flow on $\mathfrak{M}$ that projects to an equilibrium point on $\mathfrak{M}_R$ is called a *relative equilibrium.*

Example 3: Consider the equations

$$\dot{u} = v - uw(u^2 + v^2),$$

$$\dot{v} = -u - vw(u^2 + v^2),$$

$$\dot{w} = -(u^2 + v^2)^2 + (u^2 + v^2)^3,$$

where $(u, v, w) \in \mathbb{R}^3$, or in cylindrical coordinates (r, θ, w),

$$\dot{r} = -r^3 w, \quad \dot{\theta} = -1, \quad \dot{w} = -r^4 + r^6.$$

Again, these equations are independent of θ and so are invariant under rotations about the w-axis. This is the SO_2 action

$$\Phi(A, ((u, v), w) \longrightarrow (A(u, v), w).$$

(In the above, the vectors are written as row vectors but should be column vectors.) One obtains the equations on the reduced space by ignoring the θ equation. The equations on the reduced space admit the integral

$$I = \frac{1}{2}w^2 + \frac{1}{4}r^4 - \frac{1}{2}r^2,$$

and the phase portrait of this system is given in Figure 5.2. The point $r = 1, w = 0$ is an equilibrium point corresponding to a periodic solution in the full space $\mathbb{R}^3$, so it is a relative equilibrium. This equilibrium point in the reduced space is a center, and there are many periodic orbits encircling it with varying periods. If γ is one of these periodic solutions with period T, then γ is a circle in the reduced space and comes from a torus in the full space, i.e., $\Gamma = \Pi^{-1}(\gamma)$ is a torus in $\mathbb{R}^3$. If T is commensurable with 2π (the θ-period), then Γ is filled with periodic solutions, but if T is incommensurable with 2π, then Γ is filled with quasi-periodic solutions. Since the period varies from orbit to orbit on the reduced space, both phenomena occur.

An invariant set in $\mathfrak{M}$ that projects to a periodic solution on the reduced space, $\mathfrak{M}_R$, is called a *relative periodic solution*. In the third example above, the invariant torus Γ is a relative periodic solution whether it is filled with periodic solutions or quasi-periodic solutions.

A *one-parameter subgroup* of a Lie group $\mathfrak{G}$ is a closed subgroup smoothly isomorphic to $\mathbb{R}$ or $\mathbb{R}/\mathbb{Z}$, i.e., $\mathfrak{g} \subset \mathfrak{G}$ is a one-parameter subgroup if there is a smooth map $\bar{\mathfrak{g}} : \mathbb{R} \longrightarrow \mathfrak{g} \subset \mathfrak{G}$ such that $\bar{\mathfrak{g}}(t_1+t_2) = \bar{\mathfrak{g}}(t_1)\cdot\bar{\mathfrak{g}}(t_2)$ for all $t_1, t_2 \in \mathbb{R}$ and $\bar{\mathfrak{g}}(\mathbb{R}) \neq e$, e the identity in $\mathfrak{G}$. It can be shown (see [35]) that the set of all one-parameter subgroups can be put into one-to-one correspondence with tangent vectors to $\mathfrak{G}$ at the identity element. One way the correspondence is easy. Given a one-parameter subgroup $\mathfrak{g}$, the tangent vector at the identity is $d\bar{\mathfrak{g}}(t)/dt\,|_{t=0}$. The set of all such one-parameter subgroups is called the *Lie algebra of* $\mathfrak{G}$ and will be denoted by $\mathfrak{A}$; it is thought of as the tangent space at the identity of $\mathfrak{G}$. Given $A \in \mathfrak{A}$, the one-parameter subgroup is denoted $\bar{\mathfrak{g}}(t) = e^{At} = \exp(At)$.

If $\mathfrak{G}$ is one of the matrix Lie groups, then its Lie algebra is $\mathfrak{A} = \{B : e^{Bt} \in \mathfrak{G} \text{ for all } t\}$. The algebras corresponding to the standard groups given above are:

- $gl_n = gl(n, \mathbb{R})$: The algebra of all real $n \times n$ matrices.
- $o_n = o(n, \mathbb{R})$: The algebra of all real $n \times n$ skew-symmetric matrices.

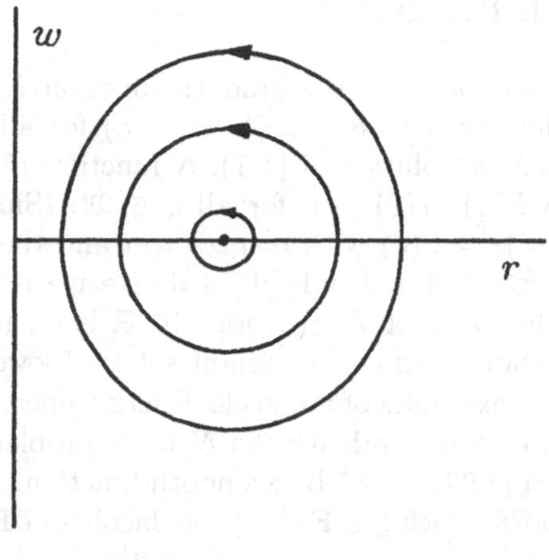

Fig. 5.2. Phase portrait of the third example

- $sl_n = sl(n, \mathbb{R})$: The algebra of all real $n \times n$ matrices with trace $= 0$.
- $so_n = so(n, \mathbb{R})$: The same as o_n.
- $sp_n = sp(n, \mathbb{R})$: The group of all real $2n \times 2n$ Hamiltonian matrices.
- $u_n = u(n, \mathbb{C})$: The group of all complex $n \times n$ skew-Hermitian matrices.

In general, an algebra is a vector space with a product. In the case of the matrix algebras given above, the product is $[A, B] = AB - BA$, the *Lie product* of two matrices.

Given an action Ψ of a group $\mathfrak{G}$ on $\mathfrak{M}$, and an element $A \in \mathfrak{A}$, one can associate a flow and hence a vector field as follows. The flow is defined by $\psi_A(t, z) = \Psi(e^{At}, z)$ and the vector field is $g_A(z) = \partial\psi_A(t, z)/\partial t \mid_{t=0}$.

For example, if $\mathfrak{G} = SO_3$, $\mathfrak{M} = \mathbb{R}^3$ with the action being matrix multiplication, then for any $A \in so_3$ the flow is $\psi_A(t, z) = e^{At}z$ and the vector field or differential equation is $\dot{z} = Az$. If $\mathfrak{G}$ with action Ψ is a symmetry group for (5.1), then by Proposition 5.1.1,

$$\psi_A(s, \phi(t, z)) = \phi(t, \psi_A(s, z)), \qquad t, s \in \mathbb{R}, \; z \in \mathfrak{M}.$$

One says that the flows ψ_A and ϕ or the vector fields g_A and f commute when the above holds. One can show

Proposition 5.1.3. *Two vector fields f and g commute if and only if $[f, g] = 0$, where $[\cdot, \cdot]$ is the Lie bracket defined by*

$$[f, g](z) = \frac{\partial f}{\partial z}(z)g(z) - \frac{\partial g}{\partial z}(z)f(z), \quad ([f, g] = Dfg - Dgf).$$

See [35] for the proof.

5.2 Systems with Integrals

An *integral* for the equation (5.1) is a smooth function $F : \mathfrak{M} \longrightarrow \mathbb{R}$ that is constant along solutions, i.e., $F(\phi(t,\zeta)) = F(\zeta)$ for all $t \in \mathbb{R}$, $\zeta \in \mathfrak{M}$, where $\phi(t,\zeta)$ is the general solution of (5.1). A function F is an integral for (5.1) if and only if $\nabla F(\zeta) \cdot f(\zeta) = 0$ for all $\zeta \in \mathfrak{M}$. (Simply differentiate the expression $F(\phi(t,\zeta)) = F(\zeta)$ with respect to t and then set $t = 0$; note that a function is constant if and only if its derivative is identically zero.) If F is an integral, then the set $F^{-1}(c)$ for $c \in \mathbb{R}$ is an invariant set, and if F is not too degenerate, then this invariant set is of lower dimension. We have already seen many examples of integrals. Energy, linear momentum, and angular momentum are all integrals for the N-body problem.

Let $\mathbf{F} = (F_1, \ldots, F_k) : \mathfrak{M} \longrightarrow \mathbb{R}^k$ be a smooth function. A *regular value of* $\mathbf{F}$ is a $c \in \mathbb{R}^k$ such that for each $\zeta \in \mathbf{F}^{-1}(c)$ the Jacobian $\partial\mathbf{F}(\zeta)/\partial\zeta = D\mathbf{F}(\zeta)$ is of maximal rank.

Proposition 5.2.1. *If c is a regular value of $\mathbf{F}$, then $\mathbf{F}^{-1}(c) \subset \mathfrak{M}$ is a smooth manifold of dimension $n - k$.*

If $k = n - 1$ and $\mathbf{F}$ is an integral for (5.1), i.e., each F_i is an integral, then $\mathbf{F}^{-1}(c)$ is a one-dimensional manifold. The only one-dimensional manifolds are unions of arcs or circles, so each component of $\mathbf{F}^{-1}(c)$ is a union of orbits of the equation, so the system is "solved."

Example: One of the classic examples of a non-Hamiltonian system with an integral is the Volteria-Lotka predator-prey system

$$\dot{u} = (a - bv)u, \qquad \dot{v} = (-c + du)v,$$

where $u, v \geq 0$ are scalar variables and a, b, c, d are positive constants. The usual fable about these equations is that u is the population of a prey and v is the population of a predator. There are two equilibrium points, $(0,0)$ and $(c/d, a/b)$. The two coordinate axes are invariant. The population of the prey u increases in the absence of predator, and the population of the predator v decreases in the absence of prey. (The origin is a saddle point.) In the first quadrant the system admits the integral

$$I = u^c e^{-du} v^a e^{-bv}.$$

The level curves of I are easily found, since it factors into a product of a function of u alone and a function of v alone. (Take logs.) The point $(c/d, a/b)$ is the point where I takes its unique global maximum value. The level curves are closed curves encircling the point $(c/d, a/b)$ (except for this point itself), representing periodic solutions of the system. So the fable says that the predator-prey populations oscillate about the equilibrium state.

Example: The Kepler problem admits the angular momentum vector $\mathbf{O} = \mathbf{q} \times \mathbf{p}$ and the energy $\mathbf{H} = \|\mathbf{p}\|^2/2 - \mu/\|q\|$ as integrals. Assume $\mathbf{O} \neq 0$ for now. The only fact used to derive this is the fact that the force is a central force. Because the force satisfies an inverse square law, there is an additional integral. From the vector identity

$$\frac{d}{dt}\left(\frac{\mathbf{q}}{\|\mathbf{q}\|}\right) = \frac{(\mathbf{q}\times\dot{\mathbf{q}})\times\mathbf{q}}{\|\mathbf{q}\|^3},$$

one gets

$$\mu\frac{d}{dt}\left(\frac{\mathbf{q}}{\|\mathbf{q}\|}\right) = \dot{p}\times\mathbf{O},$$

which integrates to

$$\mu\left\{\mathbf{e}+\frac{\mathbf{q}}{\|\mathbf{q}\|}\right\} = \mathbf{p}\times\mathbf{O}$$

where $\mathbf{e}$ is a vector integration constant called the Lorenz vector. It is an additional vector integral of the Kepler problem. Of course, not all of these integrals are independent. We have three components of $\mathbf{O}$, three components of $\mathbf{e}$, and one $\mathbf{H}$ for a total of seven integrals. However, the most you can have is five for a nontrivial system in $\mathbb{R}^6$. Clearly $\mathbf{e}\cdot\mathbf{O} = 0$, so $\mathbf{e}$ lies in the invariant plane.

It is easy to see that if $\mathbf{e} = 0$, then the motion is uniform and circular. If $\mathbf{O} \neq 0$, then by defining $\mathbf{e} = e(\cos\omega, \sin\omega)$, it is not to hard to show that ω is the argument of the perigee and e is eccentricity. Eccentricity was determined by $\mathbf{O}$ and $\mathbf{H}$, so the new integral is ω, the argument of perigee. In the planar Kepler problem, three independent integrals are $c = \|\mathbf{O}\|$, angular momentum; $h = \mathbf{H}$, energy; and ω = argument of $\mathbf{e}$, the argument of the perigee. See the lovely little book by Pollard [67].

5.3 Noether's Theorem

For Hamiltonian systems, integrals and symmetries are closely related. One implies the other.

A symplectic manifold is a differentiable manifold where Hamiltonian systems live. A *symplectic manifold* $\mathfrak{M}$ is a differentiable manifold with a special atlas of symplectic charts (or symplectic coordinates). In particular, one changes from one chart to another by a symplectic change of variables. A Hamiltonian system of equations on a symplectic manifold is a system of differential equations which is Hamiltonian in every symplectic coordinate system.

Consider the Hamiltonian system which in local symplectic coordinates z is

$$\dot{z} = J\nabla H(z). \tag{5.3}$$

(Careful: the above system of equations makes sense only in symplectic coordinates.) A function F is an integral of (5.3) if and only if $\nabla F\cdot\dot{z} = \nabla F^T J\nabla H = 0$, which leads to the definition of the Poisson bracket operator.

Let F and G be smooth functions from $\mathfrak{M}$ into $\mathbb{R}^1$ and define the *Poisson bracket* of F and G by

$$\{F,G\} = \nabla F^T J \nabla G = \frac{\partial F}{\partial u}^T \frac{\partial G}{\partial v} - \frac{\partial F}{\partial v}^T \frac{\partial G}{\partial u},$$

where $z = (u,v)$. A symplectic change of coordinates preserves the Poisson bracket, so the definition given above does not depend on the choice of symplectic coordinates z. (See [51].)

Clearly $\{F,G\}$ is a smooth map from $\mathbb{R}^{2n}$ to $\mathbb{R}$ as well, and one can easily verify that $\{\cdot,\cdot\}$ is skew-symmetric and bilinear. A little calculation verifies Jacobi's identity:

$$\{F,\{G,H\}\} + \{G,\{H,F\}\} + \{H,\{F,G\}\} = 0.$$

By the above discussion we see

Proposition 5.3.1. *F is an integral for* (5.3) *if and only if* $\{F,H\} = 0$.

By the skew-symmetry of the Poisson bracket, we see that F is an integral of the system with Hamiltonian H if and only if H is an integral of the system with Hamiltonian F.

Example: The equations of the planar Kepler problem are

$$\dot{\mathbf{q}} = \frac{\partial H}{\partial \mathbf{p}} = \mathbf{p}, \qquad \dot{\mathbf{p}} = -\frac{\partial H}{\partial \mathbf{q}} = -\frac{\mu \mathbf{q}}{\|\mathbf{q}\|^3},$$

which has angular momentum integral $\mathbf{O} = (\mathbf{q} \times \mathbf{p}) \cdot k = \mathbf{q}^T J \mathbf{p}$ as an integral. Likewise, the system

$$\dot{\mathbf{q}} = \frac{\partial \mathbf{O}}{\partial \mathbf{p}} = J\mathbf{q}, \qquad \dot{\mathbf{p}} = -\frac{\partial \mathbf{O}}{\partial \mathbf{q}} = J\mathbf{p} \tag{5.4}$$

has $H = \|p\|^2/2 - \mu/\|q\|$ as an integral. Note that the Hamiltonian flow defined by (5.4) is just rotations, i.e., $\psi(t,(\mathbf{q},\mathbf{p})) = (e^{Jt}\mathbf{q}, e^{Jt}\mathbf{p})$.

Poisson bracket predicts not only integrals for Hamiltonian systems, but also symmetries. Given functions F, G, we can form a third function $K = \{F,G\}$. Given vector fields f, g, we can form a third vector field $k = [f,g]$. They are closely related! Indeed:

Proposition 5.3.2. *If $f = J\nabla F$, $g = J\nabla G$, then $[f,g] = J\nabla\{F,G\}$.*

So the Poisson bracket is just the Lie bracket in disguise, and vice versa. Since the Lie bracket being zero means that the flows commute, the Poisson bracket being zero means the corresponding Hamiltonian flows commute also.

Proposition 5.3.3. *Let $\phi(t,\zeta)$ and $\psi(t,\zeta)$ be the Hamiltonian flows defined by $\dot{z} = J\nabla H$ and $\dot{z} = J\nabla F$ respectively. Then $\{H,F\} \equiv 0$ if and only if $\phi(t,\psi(\tau,\zeta)) \equiv \psi(\tau,\phi(t,\zeta))$ or $\phi_t \circ \psi_\tau \equiv \psi_\tau \circ \phi_t$.*

Restating what has been proved so far gives

Proposition 5.3.4. *If the system with Hamiltonian H admits an integral F, then the Hamiltonian flow defined by F is a symmetry for H, i.e., $H(\psi(\tau,\zeta)) \equiv H(\zeta)$.*

The converse is also true. A *symplectic action* is an action Ψ of a Lie group $\mathfrak{G}$ on a symplectic manifold $\mathfrak{M}$ such that the for each fixed $\gamma \in \mathfrak{G}$, the map $\Psi_\gamma = \Psi(\gamma, \cdot) : \mathfrak{M} \longrightarrow \mathfrak{M}$ is symplectic. Let $H : \mathfrak{M} \longrightarrow \mathbb{R}$ be a Hamiltonian. Then $\mathfrak{G}$ (actually the action Ψ of $\mathfrak{G}$) is a *symmetry group* of H if

$$H(\Psi(\gamma,\zeta)) \equiv H(\zeta) \quad for all \gamma \in \mathfrak{G}, \zeta \in \mathfrak{M}.$$

Then for each $A \in \mathfrak{A}$, $\psi_A(t,\zeta) = \Psi(e^{At},\zeta)$ is a Hamiltonian flow. If $\mathfrak{M}$ is simply connected, then the Hamiltonian flow ψ_A comes from a Hamiltonian system with a Hamiltonian function $F_A : \mathfrak{M} \longrightarrow \mathbb{R}$. Noether's theorem states that F_A is an integral for the Hamiltonian system defined by H.

Theorem 5.3.1. *(Noether's Theorem [61]) Let $\mathfrak{G}$ be a symmetry group of the Hamiltonian H on the simply-connected symplectic manifold $\mathfrak{M}$. Then for each element A of the Lie algebra $\mathfrak{A}$, there is an integral $F_A : \mathfrak{M} \longrightarrow \mathbb{R}$, i.e., $\{H, F_A\} \equiv 0$.*

Proof. Let $A \in \mathfrak{A}$ and $\psi_A(t,\zeta) = \Psi(e^{At},\zeta)$. Since $\psi_A(t,\zeta)$ is a Hamiltonian flow and $\mathfrak{M}$ is simply connected, $\psi_A(t,\zeta)$ is the general solution of a Hamiltonian system with Hamiltonian $F_A : \mathfrak{M} \longrightarrow \mathbb{R}$. Since $\mathfrak{G}$ is a symmetry group for H, $H(\psi_A(t,\zeta)) \equiv H(\zeta)$ or H is an integral for the F_A flow. That is, $\{F_A, H\} = 0$, but this implies F_A is an integral for the H system.

5.4 Integrals for the N-Body Problem

Consider the N-body problem in fixed coordinates with Hamiltonian

$$\mathbf{H}(\mathbf{z}) = \sum_{i=1}^{N} \frac{\|\mathbf{p}\|^2}{2m_i} - \sum_{1 \le i < j \le N} \frac{m_i m_j}{\|\mathbf{q}_i - \mathbf{q}_j\|},$$

where $\mathbf{z} = (\mathbf{q}_1, \ldots, \mathbf{q}_N, \mathbf{p}_1, \ldots, \mathbf{p}_N) \in \mathbb{R}^{6N}$. The N-body problem is defined on the symplectic manifold $\mathbb{R}^{6N} \setminus \Delta$, where Δ is the set with $\mathbf{q}_i = \mathbf{q}_j$ for some $i \neq j$. The Hamiltonian is invariant under translations, that is, the additive group $\mathbb{R}^3$ is a symmetry group for $\mathbf{H}$. The action $\Psi_T : \mathbb{R}^3 \times (\mathbb{R}^{6N} \setminus \Delta) \to \mathbb{R}^{6N}$ given by

$$\Psi_T(b, (\mathbf{q}_1, \ldots, \mathbf{q}_N, \mathbf{p}_1, \ldots, \mathbf{p}_N)) = (\mathbf{q}_1 + b, \ldots, \mathbf{q}_N + b, \mathbf{p}_1, \ldots, \mathbf{p}_N)$$

is symplectic. The algebra of $\mathbb{R}^3$ is itself and for $a \in \mathbb{R}^3$ (the algebra), the one-parameter subgroup whose tangent vector is a is

$$\psi_a(t, (\mathbf{q}_1, \dots, \mathbf{q}_N, \mathbf{p}_1, \dots, \mathbf{p}_N)) = (\mathbf{q}_1 + ta, \dots, \mathbf{q}_N + ta, \mathbf{p}_1, \dots, \mathbf{p}_N).$$

The Hamiltonian that generates ψ_a is $F = a^T(\mathbf{p}_1 + \cdots + \mathbf{p}_N)$. By Noether's theorem, $F = a^T(\mathbf{p}_1 + \cdots + \mathbf{p}_N)$ is an integral for the N-body problem for all $a \in \mathbb{R}^3$, so linear momentum $\mathbf{L} = \mathbf{p}_1 + \cdots + \mathbf{p}_N$ is an integral. In general, translational invariance implies the conservation of linear momentum.

The N-body problem is also invariant under rotation, that is, the rotation group SO_3 is a symmetry group for the N-body problem. The action $\Psi_R : SO_3 \times (\mathbb{R}^{6N} \setminus \Delta) \to \mathbb{R}^{6N} \setminus \Delta$ given by

$$\Psi_R(A, (\mathbf{q}_1, \dots, \mathbf{q}_N, \mathbf{p}_1, \dots, \mathbf{p}_N) = (A\mathbf{q}_1, \dots, A\mathbf{q}_N, A\mathbf{p}_1, \dots, A\mathbf{p}_N)$$

is symplectic and H is invariant under this action. The algebra of SO_3 is so_3, the set of all 3×3 skew-symmetric matrices. Given $B \in so_3$, we define the Hamiltonian flow

$$\psi_B(t, (\mathbf{q}_1, \dots, \mathbf{q}_N, \mathbf{p}_1, \dots, \mathbf{p}_N)) = (e^{Bt}\mathbf{q}_1, \dots, e^{Bt}\mathbf{q}_N, e^{Bt}\mathbf{p}_1, \dots, e^{Bt}\mathbf{p}_N).$$

The Hamiltonian that generates ψ_B is $F = \sum_{i=1}^N q_i^T B p_i$, so by Noether's theorem it is an integral for the N-body problem. If we take the three choices for B as follows:

$$\begin{pmatrix} 0 & 0 & 0 \\ 0 & 0 & 1 \\ 0 & -1 & 0 \end{pmatrix}, \quad \begin{pmatrix} 0 & 0 & 1 \\ 0 & 0 & 0 \\ -1 & 0 & 0 \end{pmatrix}, \quad \begin{pmatrix} 0 & 1 & 0 \\ -1 & 0 & 0 \\ 0 & 0 & 0 \end{pmatrix},$$

then the corresponding integrals are the three components of angular momentum. Therefore, the fact that the Hamiltonian is invariant under all rotations implies the law of conservation of angular momentum.

Thus the translational symmetry gives rise to three integrals, which are the three components of linear momentum, and the rotational symmetry gives rise to three integrals, the three components of angular momentum. The linear momentum integrals are always independent and the angular momentum integrals are independent unless the total angular momentum is zero. So in general there are six independent integrals (not counting H itself), so holding these integrals fixed effects a reduction from $\mathbb{R}^{6N}$ to a manifold of dimension $6N - 6$.

5.5 Symplectic Reduction

The symmetries give rise to integrals and holding the integral fixed reduces the dimension of the problem. But that is not all! Think about a problem with rotational symmetry, e.g., a problem invariant under an SO_3 action like the N-body problem discussed in the last section. We saw that this symmetry gives rise to the integrals of angular momentum. Thinking classically, angular momentum is a 3-vector pointing in space. If this vector is nonzero, the integral manifold obtained by holding it fixed is an invariant manifold of three fewer dimensions. But not all the symmetry is used up! The integral manifold is invariant under those rotations that leave the angular momentum integral fixed. In other words, there is still an SO_2 action left. Thus we can fix the three components of angular momentum and then pass to the quotient space of the SO_2 action. Since SO_2 is a one-dimensional group, the quotient space has one less dimension, so the dimension has been reduced by 3+1=4. Interestingly, this last space is symplectic and the resulting flow is Hamiltonian — see Meyer 1973 [50] and Marsden and Weinstein 1974 [44].)

As before, let $\mathfrak{M}$ be a symplectic manifold of dimension $2n$, $\Psi : \mathfrak{G} \times \mathfrak{M} \longrightarrow \mathfrak{M}$ be a symplectic action of the Lie group $\mathfrak{G}$ of dimension m, $\mathfrak{A}$ the algebra of $\mathfrak{G}$, and $H : \mathfrak{M} \longrightarrow \mathbb{R}$ a Hamiltonian that admits $\mathfrak{G}$ as a symmetry group.

Assume that there are m integrals $F_1, \dots, F_m : \mathfrak{M} \longrightarrow \mathbb{R}$. Let $\mathbf{F} = (F_1, \dots, F_m)$. Assume that $a \in \mathbb{R}^m$ is a regular value for $\mathbf{F}$ so that $\mathfrak{N} = \mathbf{F}^{-1}(a)$ is a submanifold of $\mathfrak{M}$ of dimension $2n - m$. Let $\mathfrak{G}_a$ be the subgroup of $\mathfrak{G}$ that leaves $\mathfrak{N}$ fixed. Let the dimension of $\mathfrak{G}_a$ be s. Now assume that $\mathfrak{G}_a$ acts freely and properly on $\mathfrak{N}$, so that the quotient space $\mathfrak{B} = \mathfrak{N}/\mathfrak{G}_a$ (*the reduced space*) is a manifold of dimension $2n - m - s$. The Hamiltonian H is invariant under the action, so the restriction of H to $\mathfrak{N}$ is invariant under $\mathfrak{G}_a$ and is well defined on the quotient space. Let $\bar{H} : \mathfrak{B} \longrightarrow \mathbb{R}$ be this function. Given all this we have

Theorem 5.5.1. *$\mathfrak{B}$ is a symplectic manifold. The flow defined by H on $\mathfrak{M}$ drops down to the quotient space $\mathfrak{B}$ as a Hamiltonian flow with Hamiltonian $\bar{H}$. See [50].*

This theorem is not sharp. The free and proper assumptions simply imply that the quotient space and the projection map are nice. What is essential is that the quotient and projection maps be smooth. To give a complete proof of this theorem would require a more detailed development of the theory of symplectic manifolds, so only the key idea will be given, and that may be incomprehensible to many. It may be best to skip to the next section, where the main example is discussed, or see [50, 1] for a complete proof.

The matrix J in a symplectic coordinate system defines a nondegenerate, skew-symmetric bilinear form on tangent vectors at some point $\zeta_0 \in \mathfrak{M}$. (Technically, J is the coefficients of a nondegenerate closed 2-form.) That is, if u, v are tangent vectors in a symplectic chart, we define $\{u, v\} = u^T J v$. Clearly $\{\cdot, \cdot\}$ is bilinear ($\{\alpha u_1 + \beta u_2, v\} = \alpha\{u_1, v\} + \beta\{u_2, v\}$, $\{u, \alpha v_1 + \beta v_2\} = \alpha\{u, v_1\} + \beta\{u, v_2\}, \alpha,\ \beta \in \mathbb{R}$), skew-symmetric ($\{u, v\} = -\{v, u\}$), and nondegenerate ($\{u, v\} = 0$ for all v implies $u = 0$).

This bilinear form characterizes a symplectic manifold. This bilinear form is well defined on $\mathfrak{N}$ by restriction, and by the symmetry assumption it is well defined on the quotient space $\mathfrak{B}$. What needs to be proved is that it is nondegenerate, which is established in the next lemma.

Let V be the space of all tangent vectors at the point ζ_0 and V^* its dual space. Since we are working in one symplectic coordinate system, we will identify these spaces. Let $W = \text{span}\{\nabla F_1(\zeta_0), \ldots, \nabla F_m(\zeta_0)\} \subset V$ be the linear space that is the span of all differentials to the integrals defining $\mathfrak{M}$. Thus the tangent space to $\mathfrak{M}$ is $W^0 = \{v \in V : f^T v = 0 \text{ for all } f \in W\}$. Let $W^\sharp = \{Jf : f \in W\}$, so $W^\sharp$ is the set of tangent vectors to the one-parameter group actions. That is, $W^\sharp = \{d\psi_a(0, \zeta_0)/dt,\, a \in \mathfrak{A}\}$, and $W^0 \cap W^\sharp$ is the set of tangent vectors to one-parameter groups whose orbits lie in $\mathfrak{M}$.

Thus $W^0/(W^0 \cap W^\sharp)$ is the tangent space to $\mathfrak{B}$.

Lemma 5.5.1. *If* $[u], [v] \in W^0/(W^0 \cap W^\sharp)$, *then* $\{[u], [v]\} = \{u, v\}$ *is a well-defined skew-symmetric nondegenerate bilinear form on* $W^0/(W^0 \cap W^\sharp)$.

Proof. If $\xi \in W^\sharp$, $\eta \in W^0$, then $\{\xi, \eta\} = 0$ by definition. Thus if $u, v \in W^0$ and $\xi, \eta \in W^0 \cap W^\sharp$, then $\{[u + \xi], [v + \eta]\} = \{[u + \xi], [v + \eta]\} = \{u, v\}$, so the bilinear form is well defined on the quotient space.

Now assume that $\{[u], [v]\} = 0$ for all $[v] \in W^0/(W^0 \cap W^\sharp)$. Then $\{u, v\} = 0$ for all $v \in W^0$, or $Ju \in W$. Thus $u \in W^\sharp$ or $[u] = 0$. Thus $\{\cdot, \cdot\}$ is nondegenerate on $W^0/(W^0 \cap W^\sharp)$.

This is the key lemma in the proof of the symplectic reduction theorem.

5.6 Reducing the N-Body Problem

The Hamiltonian of the N-body problem is invariant under the translation action Ψ_T, and so, as we have seen, linear momentum is a vector of integrals. Holding components of linear momentum fixed (say, equal to zero) places three linear constraints on the system, so the space where linear momentum is fixed is a $(6N - 3)$-dimensional subspace of $\mathbb{R}^{6N}$. But the action of $\mathbb{R}^3$ by Ψ_T does not change linear momentum, so all of $\mathbb{R}^3$ acts on the set where linear momentum is zero. Thus two configurations of the N bodies which are translations of one another can be identified, namely, $(\mathbf{q}_1, \ldots, \mathbf{q}_N, \mathbf{p}_1, \ldots, \mathbf{p}_N)$ and $(\mathbf{q}_1 + b, \ldots, \mathbf{q}_N + b, \mathbf{p}_1, \ldots, \mathbf{p}_N)$, where b is any vector in $\mathbb{R}^3$. Making this identification reduces the dimension by another three dimensions, making the total space $(6N - 6)$-dimensional. This space is the *first reduced space*.

The easiest way to do the reduction just discussed is to use the Jacobi coordinates given in Section 3.5. Because of later applications, we will shift the numbering system and number the particles from 0 to $N-1$. The variable $\mathbf{g}$ is the center of mass and all the other position coordinates $\mathbf{x}_1, \ldots, \mathbf{x}_{N-1}$ are relative coordinates, so the identification given above implies that $(\mathbf{g} + b, \mathbf{x}_1, \ldots, \mathbf{x}_{N-1}, \mathbf{G}, \mathbf{y}_1, \ldots, \mathbf{y}_{N-1})$ and $(\mathbf{g}, \mathbf{x}_1, \ldots, \mathbf{x}_{N-1}, \mathbf{G}, \mathbf{y}_1, \ldots, \mathbf{y}_{N-1})$ are equivalent. A representative of the equivalence class is $(0, \mathbf{x}_1, \ldots, \mathbf{x}_{N-1}, \mathbf{G}, \mathbf{y}_1, \ldots, \mathbf{y}_{N-1})$, i.e., a set with the center of mass at the origin. Linear momentum $\mathbf{G}$ is an integral, so the reduction discussed above is accomplished by setting $\mathbf{g} = 0$ and fixing $\mathbf{G}$, say to zero. The problem is described by a Hamiltonian on an even-dimensional space, the *first reduced space.* The Hamiltonian on the first reduced space is

$$H = \sum_{i=1}^{N-1} \frac{\|\mathbf{y}_i\|^2}{2M_i} - \sum_{1 \leq i < j \leq N-1} \frac{m_i m_j}{\|d_{ij}\|}.$$

Note that the problem is not Hamiltonian when just the integrals of linear momentum are fixed, but it is Hamiltonian when these integrals are fixed and points are identified by the translational symmetry.

Now consider the SO_3 symmetry given by the action Ψ_R, which gives rise to the angular momentum integrals. We work on the $(6N-6)$-dimensional first reduced space with the Jacobi coordinates $(\mathbf{x}_1, \ldots, \mathbf{x}_N, y_1, \ldots, \mathbf{y}_N)$. Recall that in Jacobi coordinates angular momentum looks the same as before, i.e.,

$$\mathbf{O} = \sum_{i=1}^{N-1} \mathbf{x}_i \times \mathbf{y}_i.$$

There are three angular momentum integrals, and they are independent except at syzygies, that is, except on configurations where the particles lie along a straight line through the center of mass. Consider the subset $\mathfrak{M} \subset \mathbb{R}^{6N-6}$ of phase space where angular momentum is some fixed, nonzero vector $\mathbf{O}$. This is a $(6N-9)$-dimensional space (submanifold), which is invariant under the flow defined by the N-body problem. Not all rotations leave $\mathfrak{M}$ fixed: only those that are rotations about $\mathbf{O}$ do. That is, let SO_2' be the subgroup of SO_3 that leaves $\mathbf{O}$ fixed. If, for example, $\mathbf{O} = c\mathbf{k}$, where $c \neq 0$ is a constant, then SO_2' comprises all matrices of the form

$$\begin{pmatrix} \cos\theta & \sin\theta & 0 \\ -\sin\theta & \cos\theta & 0 \\ 0 & 0 & 1 \end{pmatrix}.$$

So SO_2' is one-dimensional, since it can be parameterized by the angle of rotation θ.

Clearly, if $A \in SO_2'$, then A leaves $\mathfrak{M}$ invariant, so two points $z, z' \in \mathfrak{M}$ can be identified if $\Psi_R(A, z) = z'$, i.e., if one configuration can be rotated into the other by a rotation about $\mathbf{O}$. Let $\mathfrak{B}$ be the identification space $\mathfrak{M}/SO_2'$. It turns out that $\mathfrak{M}$ is $(6N-9)$-dimensional, and $\mathfrak{B}$ is $(6N-10)$-dimensional. The interesting facts are that $\mathfrak{B}$ is symplectic and the flow of the N-body problem is Hamiltonian on $\mathfrak{B}$, i.e., there are local coordinates on $\mathfrak{B}$ which are symplectic, and the equations of motion of the N-body problem are Hamiltonian.

The two reductions can be done together. The N-body problem is a first order system of differential equations in a $6N$-dimensional space $\mathbb{R}^{6N} \setminus \Delta$. The first reduction of placing the center of mass at the origin and fixing linear momentum reduces the problem to a linear subspace of dimension $6N-6$. Fixing angular momentum reduces the problem to a $(6N-9)$-dimensional space $\mathfrak{M}$. Identifying configurations which differ by a rotation about the angular momentum reduces the problem to the reduced space $\mathfrak{B}$ of dimension $6N-10$.

Consider the three-body problem in more detail. The three-body problem on $\mathfrak{B}$ is a time-independent Hamiltonian system. Two further reductions can be accomplished by holding the Hamiltonian (energy) fixed and eliminating time to get a nonautonomous system of differential equations of order 6. The reduction of the three-body problem is classical, with the elimination of the node due to Jacobi [36]. Also see [38]. These further reductions will not be needed here.

First, we will figure out the global topological type of $\mathfrak{B}$ for the three-body problem and then give a local coordinate system on $\mathfrak{B}$ which will be very useful and informative. Recall from Section 3.5 that the Hamiltonian of the three-body problem in Jacobi coordinates with the center of mass at the origin and linear momentum equal to zero is

$$H=\frac{\|\mathbf{y}_1\|^2}{2M_1}+\frac{\|\mathbf{y}_2\|^2}{2M_2}-\frac{m_0m_1}{\|\mathbf{x}_1\|}-\frac{m_1m_2}{\|\mathbf{x}_2-\alpha_0\mathbf{x}_1\|}-\frac{m_2m_0}{\|\mathbf{x}_2+\alpha_1\mathbf{x}_1\|},\tag{5.5}$$

where

$$M_1=\frac{m_0m_1}{m_0+m_1},\qquad M_2=\frac{m_2(m_0+m_1)}{m_0+m_1+m_2},$$

$$\alpha_0=\frac{m_0}{m_0+m_1},\qquad \alpha_1=\frac{m_1}{m_0+m_1}.$$

This effects the first reduction. Here, to be consistent with the later part of the book, we have labeled the masses m_0, m_1, m_2. In these coordinates, angular momentum is

$$\mathbf{O}=\mathbf{x}_1\times\mathbf{y}_1+\mathbf{x}_2\times\mathbf{y}_2.$$

Just for the fun of it, we will use Hamilton's quaternions to find the global geometry. Let $\mathbb{Q}$ denote the space of quaternions and consider phase space $(\mathbb{R}^2)^2\times(\mathbb{R}^2)^2$ as coordinatized by $\mathbb{Q}\times\mathbb{Q}$ as follows: To $(\mathbf{x}_1,\mathbf{x}_2,\mathbf{y}_1,\mathbf{y}_2)=((\mathbf{x}_1^1,\mathbf{x}_1^2),(\mathbf{x}_2^1,\mathbf{x}_2^2),(\mathbf{y}_1^1,\mathbf{y}_1^2),(\mathbf{y}_2^1,\mathbf{y}_2^2))$, associate the quaternions

$$u=\mathbf{x}_1^1+\mathbf{x}_1^2 i+\mathbf{x}_2^1 j+\mathbf{x}_2^2 k,\qquad v=\mathbf{y}_1^2+\mathbf{y}_1^1 i-\mathbf{y}_2^2 j+\mathbf{y}_2^1 k.$$

Compute $vu=\mathbf{o}+\alpha i+\beta j+\gamma k$, where $\mathbf{o}=(\mathbf{x}_1\times\mathbf{y}_1+\mathbf{x}_2\times\mathbf{y}_2)\cdot k$, $k=(0,0,1)$, is the scalar angular momentum and α,β,γ are combinations of the components of $\mathbf{x}_1,\mathbf{x}_2,\mathbf{y}_1,\mathbf{y}_2$. Thus for a given $\mathbf{o}\neq 0$, the space $\mathfrak{M}$ is

$$\mathfrak{N} = \{(u,v) \in \mathbb{Q} \times \mathbb{Q} : u \neq 0 \text{ and } v = (o + \alpha i + \beta j + \gamma k)u^{-1}\}.$$

Thus $\mathfrak{N}$ is coordinatized by $u \in \mathbb{Q} \setminus \{0\} \simeq S^3 \times \mathbb{R}^1$ and $(\alpha, \beta, \gamma) \in \mathbb{R}^3$, or $\mathfrak{N}$ is just $S^3 \times \mathbb{R}^4$.

The SO_2 action on $(\mathbb{R}^2)^2 \times (\mathbb{R}^2)^2$ is equivalent to the S^1 action on $\mathbb{Q} \times \mathbb{Q}$ given by $(\theta, (u,v)) \longrightarrow (r(\theta)u, vr(\theta)^{-1})$, where $\theta \in S^1$ and $r(\theta) = \cos\theta + i \sin\theta$. Thus to pass down to $\mathfrak{B} = \mathfrak{N}/SO_2$ is to identify the points

$$(u, \{(o + \alpha i + \beta j + \gamma k\}u^{-1}) \text{ and } (r(\theta)u, \{(o + \alpha i + \beta j + \gamma k\}(r(\theta)u)^{-1})$$

on $\mathfrak{N}$. Note that the identified points both have the same coordinates in the last three places, namely α, β, γ. Thus $\mathfrak{B} = \{(\mathbb{Q} \setminus \{0\})/SO_2)\} \times \mathbb{R}^3$. For each three-sphere about $\{0\} \in \mathbb{Q}$, the SO_2 action is just $(\theta, u) \longrightarrow r(\theta)u$, which is the usual action giving rise to the Hopf fibration, so $(\mathbb{Q} \setminus \{0\})/SO_2 \simeq S^2 \times \mathbb{R}^1$. Thus $\mathfrak{B} = S^2 \times \mathbb{R}^4$. This result is found in [24], but the quaternions calculation came from [50].

Now let us construct local coordinates on $\mathfrak{B}$. Putting the Hamiltonian (5.5) in polar coordinates gives

$$\begin{aligned} H = {} & \frac{1}{2M_1}\left\{R_1^2 + \left(\frac{\Theta_1^2}{r_1^2}\right)\right\} + \frac{1}{2M_2}\left\{R_2^2 + \left(\frac{\Theta_2^2}{r_2^2}\right)\right\} - \frac{m_0 m_1}{r_1} \\ & - \frac{m_0 m_2}{\sqrt{r_2^2 + \alpha_0^2 r_1^2 - 2\alpha_0 r_1 r_2 \cos(\theta_2 - \theta_1)}} \\ & - \frac{m_1 m_2}{\sqrt{r_2^2 + \alpha_1^2 r_1^2 + 2\alpha_1 r_1 r_2 \cos(\theta_2 - \theta_1)}}. \end{aligned}$$

Since the Hamiltonian depends only on the difference of the two polar angles, make the symplectic change of coordinates

$$\phi_1 = \theta_1, \qquad \phi_2 = \theta_2 - \theta_1,$$

$$\Phi_1 = \Theta_1 + \Theta_2, \ \Phi_2 = \Theta_2.$$

Since the Hamiltonian is independent of ϕ_1, it is an ignorable coordinate and its conjugate Φ_1, total angular momentum, is a constant. The reduction is accomplished by ignoring ϕ_1 and setting $\Phi_1 = c$, where c is a constant. The Hamiltonian on the reduced space becomes

$$\begin{aligned} H = {} & \frac{1}{2M_1}\left\{R_1^2 + \left(\frac{(c - \Phi_2)^2}{r_1^2}\right)\right\} + \frac{1}{2M_2}\left\{R_2^2 + \left(\frac{\Phi_2^2}{r_2^2}\right)\right\} - \frac{m_0 m_1}{r_1} \\ & - \frac{m_1 m_2}{\sqrt{r_2^2 + \alpha_0^2 r_1^2 + 2\alpha_0 r_1 r_2 \cos(\phi_2)}} \\ & - \frac{m_2 m_0}{\sqrt{r_2^2 + \alpha_1^2 r_1^2 - 2\alpha_1 r_1 r_2 \cos(\phi_2)}}. \end{aligned} \tag{5.6}$$

The Hamiltonian (5.6) has three degrees of freedom with symplectic coordinates $(r_1, r_2, \phi_2, R_1, R_2, \Phi_2)$ and one parameter c. Thus we have the Hamiltonian of the three-body problem in local coordinates.

5.7 Problems

1 Verify that GL_n, O_n, SL_n,SO_n, Sp_n are groups.

2 Verify
- $A \in sl_n$ if and only if $e^{At} \in SL_n$ for all t.
- $A \in so_n$ if and only if $e^{At} \in SO_n$ for all t.
- $A \in sp_n$ if and only if $e^{At} \in Sp_n$ for all t.

3 A *Lie algebra* is a vector space $\mathcal{A}$ with a non-associative product $[\cdot, \cdot] : \mathcal{A} \times \mathcal{A} \to \mathcal{A}$ which is linear in both arguments (bilinear) and satisfies the Jacobi identity

$$[A, [B, C]] + [B, [C, A]] + [C, [A, B]] = 0.$$

verify that gl_n, sl_n, so_n, sp_n are Lie algebras when the product is $[A, B] = AB - BA$.

4 A system of equations $\dot{u} = f(u)$, $u \in \mathbb{R}^{2n}$ admits a *time reversing symmetry* or is *reversible* if $f(Ru) = -Rf(u)$ where R is an $2n \times 2n$ matrix such that R is similar to $diag\,\{I_n, -I_n\}$. So $R^2 = I_{2n}$. Show that if $\phi(t)$ is a solution then so is $R\phi(-t)$.

5 Write $\ddot{u} + u^3 = 0$ as a system and show that $R = diag\,\{1, -1\}$ makes the system reversible. What about $R' = diag\,\{-1, 1\}$?

6 Let $H(u)$ be a Hamiltonian such that $H(Su) = H(u)$ where S is an $2n \times 2n$ matrix such that S is anti-symplectic ($S^T JS = -J$) and S is similar to $diag\,\{I_n, -I_n\}$. Show that the system defined by H is reversible. In this case we say that H defines a reversible Hamiltonian system.

7 Show that a classical Hamiltonian system of the form $H(q, p) = p^T Mp + V(q)$ defines a reversible Hamiltonian system where M is an $n \times n$ symmetric matrix and $V : \mathbb{R}^n \to \mathbb{R}$.

8 Show that the restricted problem (2.7) admits a time reversing symmetry. (Hint: $S = diag\,(1, -1, -1, 1)$.)

6. Theory of Periodic Solutions

In this chapter, the basic theory of periodic solutions, their continuation, and their stability is presented. The first two topics are very closely related, since many of the questions about equilibrium points are very similar to questions about fixed points. Later, we will show that periodic solutions are related to both.

6.1 Equilibrium Points

Consider first a general system

$$\dot{z} = f(z), \tag{6.1}$$

where $f : \mathcal{O} \to \mathbb{R}^n$ is smooth and $\mathcal{O}$ is open in $\mathbb{R}^n$. The results in this section are of a local nature and so we can work in one coordinate system, z. Let the general solution be $\phi(t, \zeta)$, i.e., $\phi(t, \zeta)$ is the solution of (6.1) such that $\phi(0, \zeta) = \zeta$. An *equilibrium point (rest point, critical point, stationary point)* is a $z_0 \in \mathcal{O}$ such that $f(z_0) = 0$. It gives rise to an equilibrium solution $\phi(t, z_0)$ such that $\phi(t, z_0) \equiv z_0$ for all t: so questions about the existence and uniqueness of equilibrium solutions are finite-dimensional questions. The eigenvalues of $\partial f(z_0)/\partial z$ are called the *(characteristic) exponents* of the equilibrium point. If $\partial f(z_0)/\partial z$ is nonsingular, or equivalently the exponents are all nonzero, then the equilibrium point is called *elementary.*

Proposition 6.1.1. *Elementary equilibrium points are isolated.*

Proof. $f(z_0) = 0$ and $\partial f(z_0)/\partial z$ is nonsingular, so the inverse function theorem applies to f; therefore, there is a neighborhood of z_0 with no other zeros of f.

Henceforth, let the equilibrium point be at the origin, i.e., $z_0 = 0$. The analysis of stability, bifurcations, etc. of equilibrium points starts with an analysis of the linearized equations. For this reason, one rewrites (6.1) as

$$\dot{z} = Az + g(z), \tag{6.2}$$

where $A = \partial f(0)/\partial z$, $g(z) = f(z) - Az$; so $g(0) = 0$ and $\partial g(0)/\partial z = 0$. The nonlinear terms are contained in g.

The *linearized equations (about z_0)* are obtained by setting $g(z) = 0$. The eigenvalues of A are the exponents of the equilibrium point, so called because the linearized equations (e.g., $g(z) \equiv 0$ in (6.2)) have solutions which contain terms like $\exp(\lambda t)$, where λ is an eigenvalue of A.

There are many different stability concepts. Here are a few:

- The equilibrium point $z = 0$ is said to be *positively (negatively) stable* if for every $\varepsilon > 0$ there is a $\delta > 0$ such that $\|\phi(t,\zeta)\| < \varepsilon$ for all $t \geq 0$ ($t \leq 0$) whenever $\|\zeta\| < \delta$.
- The equilibrium point $z = 0$ is said to be *stable* if it is both positively and negatively stable.
- The equilibrium $z = 0$ is *unstable* if it is not stable. (The adjectives "positively" and "negatively" can be used with "unstable" also.)
- The equilibrium $z = 0$ is *asymptotically stable* if it is positively stable and there is an $\eta > 0$ such that $\phi(t,\zeta) \longrightarrow 0$ as $t \longrightarrow +\infty$ for all $\|\zeta\| < \eta$.
- The equilibrium is *spectrally stable* if all its exponents are pure imaginary. (In Hamiltonian systems, the equilibrium is called *elliptic.*)
- The equilibrium is *linearly stable* if it is spectrally stable and the matrix A is diagonalizable.

In many books "stable" means positively stable, but the above convention is the common one in the theory of Hamiltonian differential equations. If all the exponents have negative real parts, then a classical theorem of Liapunov states that the origin is asymptotically stable; see [18, 31, 39]. But the eigenvalues of a Hamiltonian matrix are symmetric with respect to the imaginary axis, so this theorem never applies to Hamiltonian systems — see [51]. In fact, since the flow defined by a Hamiltonian system is volume-preserving, an equilibrium point can never be asymptotically stable.

Liapunov also proved that if one exponent has a positive real part, then the origin is positively unstable [18, 31, 39]. Thus a necessary condition for the stability of the origin is that all the eigenvalues be pure imaginary, whence the definition of spectrally stable. However, linear stability does not imply stability — see [51]. If all the exponents have real parts different from zero, then the equilibrium point is called *hyperbolic*.

Assume now that the differential equations depend on some parameters. Consider

$$\dot{z} = f(z, \nu), \tag{6.3}$$

where $f : \mathcal{O} \times \mathcal{Q} \longrightarrow \mathbb{R}^n$ is smooth, $\mathcal{O}$ is open in $\mathbb{R}^n$, and $\mathcal{Q}$ is open in $\mathbb{R}^k$. The general solution $\phi(t, z, \nu)$ is smooth in the parameter ν also.

Let $z = z_0$ be an equilibrium point when $\nu = \nu^*$ (i.e., $f(z_0, \nu^*) = 0$). A *continuation* of this equilibrium point is a smooth function $\eta(\nu)$ defined for ν near ν^* such that $\eta(\nu^*) = z_0$ and $\eta(\nu)$ is an equilibrium point for all ν near ν^* (i.e., $f(\eta(\nu), \nu) = 0$).

Proposition 6.1.2. *An elementary equilibrium point can be continued.*

Proof. Apply the implicit function theorem to $f(z,\nu) = 0$. By assumption, $f(z_0,\nu^*) = 0$ and $\partial f(z_0,\nu^*)/\partial z$ is nonsingular; so the implicit function theorem asserts the existence of the function $\eta(\nu)$ such that $\eta(\nu^*) = z_0$ and $f(\eta(\nu),\nu) \equiv 0$.

Corollary 6.1.1. *The exponents of elementary equilibrium points vary continuously with the parameter* ν.

Proof. The exponents of the equilibrium $\eta(\nu)$ are the eigenvalues of the Jacobian $\partial f(\eta(\nu),\nu)/\partial z$. This matrix varies smoothly with the parameter ν, so its eigenvalues vary continuously with the parameter ν. The dependence may not be differentiable.

6.2 Fixed Points

Consider a diffeomorphism

$$z \longrightarrow z' = f(z), \tag{6.4}$$

where $f : \mathcal{O} \to \mathbb{R}^n$ is smooth and $\mathcal{O}$ is open in $\mathbb{R}^n$. We think of f as defining a discrete dynamical system, i.e., the orbit of a point z is $\bigcup_{-\infty}^{+\infty} f^k(z)$, where f^k is the kth iterate of f and

$$f^k = f \circ f \circ \cdots \circ f \qquad k \text{ times for } k > 0,$$

$$f^0 \text{ is the identity map,}$$

$$f^{-k} = f^{-1} \circ f^{-1} \circ \cdots \circ f^{-1} \qquad k \text{ times for } k > 0.$$

A *fixed point* is a $z_0 \in \mathbb{R}^n$ such that $f(z_0) = z_0$, or $f(z_0) - z_0 = 0$: questions about the existence and uniqueness of fixed points are finite-dimensional questions. The eigenvalues of $\partial f(z_0)/\partial z$ are called the *(characteristic) multipliers* of the fixed point. If $\partial f(z_0)/\partial z - I$ is nonsingular, or equivalently, the multipliers are all different from $+1$, then the fixed point is called *elementary.*

Proposition 6.2.1. *Elementary fixed points are isolated.*

Proof. Apply the implicit function theorem to $h(z) = f(z) - z = 0$. Since $h(z_0) = 0$ and $\partial h(z_0)/\partial z = \partial f(z_0)/\partial z - I$ is nonsingular, the implicit function theorem applies to h. Therefore, there is a neighborhood of z_0 with no other zeros of h or fixed points of f.

Henceforth, let the fixed point be at the origin, i.e., $z_0 = 0$. The analysis of stability, bifurcations, etc. of fixed points starts with an analysis of the linearized equations. For this reason, one rewrites (6.4) as

$$z \longrightarrow z' = Az + g(z), \tag{6.5}$$

where $A = \partial f(0)/\partial z$, $g(z) = f(z) - Az$, so $g(0) = 0$ and $\partial g(0)/\partial z = 0$. The eigenvalues of A are the multipliers of the fixed point. The *linearized map at* z_0 is obtained by setting $g = 0$.

Again there are many different stability concepts. Here are a few:

- The fixed point $z = 0$ of (6.4) is said to be *positively (negatively) stable* if for every $\varepsilon > 0$ there is a $\delta > 0$ such that $\|f^k(z)\| < \varepsilon$ for all $k \geq 0$ ($k \leq 0$) whenever $\|z\| < \delta$.
- The fixed point $z = 0$ is said to be *stable* if it is both positively and negatively stable.
- The fixed point $z = 0$ is *unstable* if it is not stable. (The adjectives "positively" and "negatively" can be used with "unstable" also.)
- The fixed point $z = 0$ is *asymptotically stable* if it is positively stable and there is an $\eta > 0$ such that $f^k(z) \longrightarrow 0$ as $k \longrightarrow +\infty$ for all $\|z\| < \eta$.
- The fixed point is *spectrally stable* if all its multipliers have absolute value 1. (For symplectic maps, the fixed point is called *elliptic*.)
- The fixed point is *linearly stable* if it is spectrally stable and the matrix A is diagonalizable.

If all the multipliers have absolute value less than 1, then a classical theorem states that the origin is asymptotically stable; see [18, 31]. But the eigenvalues of a symplectic matrix are symmetric with respect to the unit circle, so this theorem never applies to symplectic maps — see [51]. In fact, since the diffeomorphism is volume-preserving, a fixed point can never be asymptotically stable.

Also, if one multiplier has absolute value greater than 1, then the origin is positively unstable [18, 31]. Thus a necessary condition for stability of the origin is that all the eigenvalues have absolute value 1. However, linear stability does not imply stability. If all the multipliers have absolute values different from 1, then the fixed point is called *hyperbolic*.

Assume now that the diffeomorphism depends on some parameters; consider

$$z \longrightarrow z' = f(z, \nu), \tag{6.6}$$

where $f : \mathcal{O} \times \mathcal{Q} \longrightarrow \mathbb{R}^n$ is smooth, $\mathcal{O}$ is open in $\mathbb{R}^n$, and $\mathcal{Q}$ is open in $\mathbb{R}^k$. $f^k(z, \nu)$ is smooth in the parameter ν also.

Let $z = z_0$ be a fixed point when $\nu = \nu^*$ (i.e., $f(z_0, \nu^*) = z_0$). A *continuation* of this fixed point is a smooth function $\eta(\nu)$ defined for ν near ν^* such that $\eta(\nu^*) = z_0$ and $\eta(\nu)$ is a fixed point for all ν near ν^* (i.e., $f(\eta(\nu), \nu) = \eta(\nu)$).

Proposition 6.2.2. *An elementary fixed point can be continued and the multipliers vary continuously with the parameter* ν.

Proof. Apply the implicit function theorem to $h(z,\nu) = f(z,\nu) - z = 0$.

6.3 Periodic Differential Equations

Consider the periodic system

$$\dot{z} = f(t,z), \tag{6.7}$$

where $f : \mathbb{R} \times \mathcal{O} \to \mathbb{R}^n$ is smooth and $\mathcal{O}$ is open in $\mathbb{R}^n$. Let f be T-periodic in t with $T > 0$, i.e., $f(t+T,z) = f(t,z)$ for all $(t,z) \in \mathbb{R} \times \mathcal{O}$. Let the general solution be $\phi(t,\zeta)$, i.e., $\phi(t,\zeta)$ is the solution of (6.7) such that $\phi(0,\zeta) = \zeta$. A periodic solution of (6.7) is a solution $\phi(t,\zeta_0)$ such that $\phi(t+T,\zeta_0) \equiv \phi(t,\zeta_0)$ for all t.

Lemma 6.3.1. *A necessary and sufficient condition for* $\phi(t,\zeta_0)$ *to be periodic with a period* T *is*

$$\phi(T,\zeta_0) = \zeta_0. \tag{6.8}$$

Proof. Let $\psi(t) = \phi(t+T,\zeta_0)$. By assumption $\psi(0) = \zeta_0$ and $\dot{\psi}(t) = \dot{\phi}(t+T,\zeta_0) = f(t+T,\phi(t+T,\zeta_0)) = f(t,\phi(t+T,\zeta_0)) = f(t,\psi(t))$. Thus $\psi(t)$ satisfies the same equation and initial conditions, so the uniqueness theorem for ordinary differential equations implies $\phi(t,\zeta_0) = \psi(t) = \phi(t+T,\zeta_0)$.

This lemma shows that questions about the existence and uniqueness of periodic solutions are ultimately finite-dimensional questions. The analysis and topology of finite-dimensional spaces should be enough to answer all such questions.

We will reduce all questions about periodic solutions to questions about diffeomorphisms. For the periodic system (6.7), define the period map P to be

$$P(z) = \phi(T,z),$$

so the map $z \longrightarrow z' = P(z)$ is a diffeomorphism. By the lemma above, a point ζ_0 is the initial condition for a T-periodic solution if and only if it is a fixed point of P.

Let $\phi(t,\zeta_0)$ be a periodic solution. The matrix $\partial\phi(T,\zeta_0)/\partial z$ is called the *monodromy matrix*, and its eigenvalues are called the *(characteristic) multipliers* of the periodic solution. Note the multipliers are the same as the multipliers of the corresponding fixed point of the period map. We will say that the periodic solution is "stable" ("linearly stable", etc.) if the corresponding fixed point is stable (linearly stable, etc.)

Example: A simple example to illustrate these ideas is the forced Duffing's equation. One version of Duffing's equation is

$$\ddot{u} + \omega^2 u + \gamma u^3 = A \cos t.$$

Assume that the forcing is small by setting $A = \varepsilon B$ and treating ε as a small parameter. Writing this as a system, we have

$$\begin{pmatrix} \dot{u} \\ \dot{v} \end{pmatrix} = \begin{pmatrix} \omega v \\ -\omega u - (\gamma/\omega)u^3 + \varepsilon(B/\omega)\cos t \end{pmatrix}.$$

When $\varepsilon = 0$, the system has a 2π periodic solution $u = v = 0$. The linear variational equation about this solution is

$$\begin{pmatrix} \dot{u} \\ \dot{v} \end{pmatrix} = \begin{pmatrix} 0 & \omega \\ -\omega & 0 \end{pmatrix} \begin{pmatrix} v \\ u \end{pmatrix}.$$

The fundamental matrix solution is

$$e^{At} = \begin{pmatrix} \cos \omega t & \sin \omega t \\ \\ -\sin \omega t & \cos \omega t \end{pmatrix}.$$

Now $e^{A2\pi} - I$ is nonsingular if and only if $\omega \neq n$ with $n \in \mathbb{Z}$. That is, if the natural frequency ω is not an integral multiple of the forcing frequency 1. In this case, the theory above says that Duffing's equation has a small 2π-periodic solution.

The main example of a periodic system treated in this book is the elliptic restricted three-body problem discussed in Chapter 11. All the other examples are autonomous (time-independent).

6.4 Autonomous Systems

Consider again a general autonomous system

$$\dot{z} = f(z), \tag{6.9}$$

where $f : \mathcal{O} \to \mathbb{R}^n$ is smooth and $\mathcal{O}$ is open in $\mathbb{R}^n$. Let the general solution be $\phi(t, \zeta)$. As above, a solution $\phi(t, \zeta_0)$ is T-periodic, $T > 0$, if and only if

$$\phi(T, \zeta_0) = \zeta_0. \tag{6.10}$$

There is a problem already, since the period T is not defined by the equation. There is no external clock!

It is tempting to use the implicit function theorem on (6.10) to find a condition for local uniqueness of a periodic solution. To apply the implicit function theorem to (6.10), the matrix $\partial\phi(T, \zeta_0)/\partial z - I$ would have to be nonsingular, or equivalently, 1 would not be a multiplier. But this will never happen.

Lemma 6.4.1. *Periodic solutions of* (6.9) *are never isolated, and* +1 *is always a multiplier. In fact,* $f(\zeta_0)$ *is an eigenvector of the monodromy matrix corresponding to the eigenvalue* +1.

Proof. Since (6.9) is autonomous, it defines a local dynamical system, so a time-translate of a solution is a solution. Therefore, the periodic solution is not isolated. Differentiating the group relation $\phi(\tau, \phi(t, \zeta_0)) = \phi(t + \tau, \zeta_0)$ with respect to t and setting $t = 0$ and $\tau = T$, we have

$$\frac{\partial \phi}{\partial z}(T, \zeta_0)\dot{\phi}(0, \zeta_0) = \dot{\phi}(T, \zeta_0),$$

$$\frac{\partial \phi}{\partial z}(T, \zeta_0)f(\zeta_0) = f(\zeta_0).$$

Since the periodic solution is not an equilibrium point, $f(\zeta_0) \neq 0$.

Because of this lemma, the correct concept is "isolated periodic orbit." In order to overcome the difficulties implicit in this lemma, one introduces a cross section. Let $\phi(t, \zeta_0)$ be a periodic solution. A *cross section to the periodic solution*, or simply *a section*, is a hyperplane Σ of codimension 1 through ζ_0 and transverse to $f(\zeta_0)$. For example, Σ would be the hyperplane $\{z : a^T(z - \zeta_0) = 0\}$, where a is a constant vector with $a^T f(\zeta_0) \neq 0$. The periodic solution starts on the section and, after a time T, returns to it. By the continuity of solutions with respect to initial conditions, nearby solutions do the same. See Figure 6.1. So if z is close to ζ_0 on Σ, there is a time $\mathcal{T}(z)$ close to T such that $\phi(\mathcal{T}(z), z)$ is on Σ. $\mathcal{T}(z)$ is called the *first return time.* The *section map*, or *Poincaré map*, is defined as the map $P : z \longrightarrow \phi(\mathcal{T}(z), z)$, which is a map from a neighborhood N of ζ_0 in Σ into Σ.

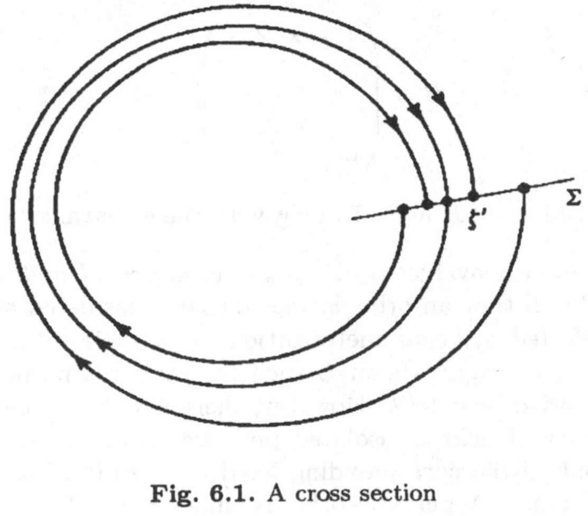

Fig. 6.1. A cross section

Lemma 6.4.2. *If the neighborhood N of ζ_0 in Σ is sufficiently small, then the first return time, $\mathcal{T} : N \longrightarrow \mathbb{R}$, and the Poincaré map, $P : N \longrightarrow \Sigma$, are smooth.*

Proof. Let $\Sigma = \{z : a^T(z - \zeta_0) = 0\}$, where $a^T f(\zeta_0) \neq 0$. Consider the function $g(t, z) = a^T(\phi(t, \zeta) - \zeta_0)$. Since $g(T, \zeta_0) = 0$ and $\partial g(T, \zeta_0)/\partial t = a^T \dot{\phi}(T, \zeta_0) = a^T f(\zeta_0) \neq 0$, the implicit function theorem gives a smooth function $\mathcal{T}(z)$ such that $g(\mathcal{T}(z), z) = 0$. If g is zero, then it defines Σ so that the first return time $\mathcal{T}$ is smooth. The Poincaré map is smooth because it is the composition of two smooth maps.

The periodic solution now appears as a fixed point of P; indeed, any fixed point $z^\dagger$ of P is the initial condition for a periodic solution of period $\mathcal{T}(z^\dagger)$, since $(\mathcal{T}(z^\dagger), z^\dagger)$ would satisfy (6.8). A point $z^\dagger \in N$ such that $P^k(z^\dagger) = z^\dagger$ for some integer $k > 0$ is called a *periodic point* of P of period k. The solution of (6.9) through such a periodic point will be periodic with period approximately kT.

The definitions of monodromy matrix and multipliers are the same as for periodic systems. (Indeed, an autonomous system is a T-periodic system for any T.)

Lemma 6.4.3. *If the multipliers of the periodic solution are $1, \lambda_2, \ldots, \lambda_n$, then the multipliers of the corresponding fixed point of the Poincaré map are $\lambda_2, \ldots, \lambda_n$.*

Proof. First translate coordinates so that $\zeta_0 = 0$ and then rotate the coordinates so that $f(\zeta_0) = (1, 0, \ldots, 0)$, so Σ is the hyperplane $z_1 = 0$. Let $B = \partial\phi(T, \zeta_0)/\partial z$, the monodromy matrix. By Lemma 6.4.1, $f(\zeta_0)$ is an eigenvector of B corresponding to the eigenvalue $+1$. In these coordinates,

$$B = \begin{pmatrix} 1 & \times & \times & \times & \times \\ 0 & & & & \\ \vdots & & A & & \\ 0 & & & & \end{pmatrix}$$

Clearly the eigenvalues of B are $+1$ along with the eigenvalues of A.

We also call the eigenvalues $\lambda_2, \ldots, \lambda_n$ the *nontrivial multipliers* of the periodic orbit. Recall that an orbit is the solution considered as a curve in $\mathbb{R}^n$, so it is unaffected by reparameterization. A periodic orbit of period T is *isolated* if it has a neighborhood L such that there is no other periodic orbit in L with period near to T. However, there may be periodic solutions of much higher period near an isolated periodic orbit. A periodic orbit is isolated if and only if the corresponding fixed point of the Poincaré map is an isolated fixed point. A periodic orbit is called *elementary* if none of its nontrivial multipliers is $+1$. As before, we have:

Proposition 6.4.1. *Elementary periodic orbits are isolated and can be continued.*

We shall say that a periodic solution is stable, spectrally stable, etc. if the corresponding fixed point of the Poincaré map has the same property. But be careful: the timing is lost and this definition of stable is not the usual definition due to Liapunov. The usual definition is as follows: the periodic solution $\phi(t, \zeta_0)$ is ***Liapunov stable*** if for each $\varepsilon > 0$ there is a $\delta > 0$ such that $\|\phi(t, \zeta) - \phi(t, \zeta_0)\| < \varepsilon$ for all $t \in \mathbb{R}$ provided $\|\zeta - \zeta_0\| < \delta$.

The definition used here is usually called orbital stability. Let $\phi(t, \zeta_0)$ be a periodic solution and let $\mathcal{P} = \{\phi(t, \zeta_0) : t \in \mathbb{R}\}$, so $\mathcal{P}$ is the orbit of the periodic solution. The periodic solution $\phi(t, \zeta_0)$ is ***orbitally stable*** if for each $\varepsilon > 0$, there is a $\delta > 0$ such that $d(\phi(t, \zeta), \mathcal{P}) < \varepsilon$ for all $t \in \mathbb{R}$, provided $\|\zeta - \zeta_0\| < \delta$. Here d is the distance from a point to a set.

Example: Consider the system

$$\dot{u} = v(1 + u^2 + v^2), \qquad \dot{v} = -u(1 + u^2 + v^2),$$

or, in polar coordinates,

$$\dot{r} = 0, \qquad \dot{\theta} = -1 - r^2.$$

All solutions are periodic, but the periods vary. Each orbit is a circle centered at the origin. If two points are close, then the circles they lie on are uniformly close. But the angular velocity differs in each circle, so if two solutions start near to each other but on different circles, then at some time they will be far apart. These solutions are orbitally stable (or stable in our sense), but not Liapunov stable. To see that these solutions are stable in our sense, note that $\theta = 0, r > 0$ is a cross section for all the non-equilibrium solutions and the Poincaré map is the identity map.

Example: Consider the system

$$\dot{u} = v + u(1 - u^2 - v^2), \qquad \dot{v} = -u + v(1 - u^2 - v^2),$$

which in polar coordinates is

$$\dot{r} = r(1 - r^2), \qquad \dot{\theta} = -1.$$

The origin is an elementary equilibrium point, and the unit circle is an elementary periodic orbit. To see the latter claim, consider the cross section $\theta \equiv 0 \bmod 2\pi$. The first return time is 2π. The linearized equation about $r = 1$ is $\dot{r} = -2r$, so the linearized Poincaré map is $r \longrightarrow r\exp(-4\pi)$. The multiplier of the fixed point is $\exp(-4\pi)$.

6.5 Systems with Integrals

As we have seen, the monodromy matrix of a periodic solution has $+1$ as a multiplier. If equation (6.9) were Hamiltonian, the monodromy matrix would be symplectic, so the algebraic multiplicity of the eigenvalue +1 would be even and hence at least 2. Actually, this is simply due to the fact that an autonomous Hamiltonian system has an integral.

Throughout this section, assume that equation (6.9) admits an integral F, where F is a smooth map from $\mathcal{O}$ to $\mathbb{R}$, and assume that $\phi(t, \zeta_0)$ is a periodic solution of period T. Furthermore, assume that the integral F is nondegenerate on this periodic solution, i.e., $\nabla F(\zeta_0)$ is nonzero. For a Hamiltonian system, the Hamiltonian H is always nondegenerate on a non-equilibrium solution since $\nabla H(\zeta_0) = 0$ would imply an equilibrium.

Lemma 6.5.1. *If F is nondegenerate on the periodic solution $\phi(t, \zeta_0)$, then the multiplier +1 has algebraic multiplicity of at least 2. Moreover, the row vector $\partial F(\zeta_0)/\partial x$ is a left eigenvector of the monodromy matrix corresponding to the eigenvalue +1.*

Proof. Differentiating $F(\phi(t, \zeta)) \equiv F(z)$ with respect to z and setting $z = \zeta_0$ and $t = T$, we have

$$\frac{\partial F(\zeta_0)}{\partial z} \frac{\partial \phi(T, \zeta_0)}{\partial z} = \frac{\partial F(\zeta_0)}{\partial z},$$

which implies the second part of the lemma. Choose coordinates so that $f(\zeta_0)$ is the column vector $(1, 0, \ldots, 0)^T$ and $\partial F(\zeta_0)/\partial z$ is the row vector $(0, 1, 0, \ldots, 0)$. Since $f(\zeta_0)$ is a right eigenvector and $\partial F(\zeta_0 \partial z$ is a left eigenvector, the monodromy matrix $B = \partial \phi(T, \zeta_0)/\partial z$ has the form

$$B = \begin{pmatrix} 1 & \times & \times & \times & \times \\ 0 & 1 & 0 & 0 & 0 \\ 0 & \times & \times & \times & \times \\ 0 & \times & \times & \times & \times \\ \vdots & & & & \\ 0 & \times & \times & \times & \times \end{pmatrix}$$

Expand by minors and let $p(\lambda) = \det(B - \lambda I)$. First, expand along the first column to get $p(\lambda) = (1 - \lambda)\det(B' - \lambda I)$, where B' is the $(m-1) \times (m-1)$ matrix obtained from B by deleting the first row and column. Next, expand $\det(B' - \lambda I)$ along the first row to get $p(\lambda) = (1 - \lambda)^2 \det(B'' - \lambda I) = (1 - \lambda)^2 q(\lambda)$, where B'' is the $(m-2) \times (m-2)$ matrix obtained from B by deleting the first two rows and columns.

Again, there is a good geometric reason for the degeneracy implied by this lemma. The periodic solution lies in an $(m-1)$-dimensional level set of the integral, and typically in nearby level sets of the integral, there is a periodic orbit. So periodic orbits are not isolated.

Consider the Poincaré map $P : N \longrightarrow \Sigma$, where N is a neighborhood of w' in Σ. Let ξ be flow box coordinates at w', that is, ξ is a local coordinate system at w' with w' corresponding to $\xi = 0$, and the equations (6.9) in these coordinates are $\dot{\xi}_1 = 1, \dot{\xi}_2 = 0, \ldots, \dot{\xi}_n = 0$, and $F(\xi) = \xi_2$ — see [51]. In these coordinates, we may take Σ to be $\xi_1 = 0$. Since ξ_2 is the integral in these coordinates, P maps the level sets $\xi_2 = constant$ into themselves, so we can ignore the ξ_2 component of P. Let $e = \xi_2$, let Σ_e be the intersection of Σ and the level set $F = e$, and let $\xi_3, \ldots, \xi_n$ be coordinates in Σ_e. Here e is considered as a parameter (the value of the integral). In these coordinates, the Poincaré map P is a function of $\zeta = (\xi_3, \ldots, \xi_n)$ and the parameter e. So $P(e, \zeta) = (e, Q(e, \zeta))$, where for fixed $e, Q(e, \cdot)$ is a mapping of a neighborhood N_e of the origin in Σ_e into Σ_e. Q is called the *Poincaré map in an integral surface.* The eigenvalues of $\partial Q(0,0)/\partial\zeta$ are called the *multipliers of the fixed point in the integral surface* or the *nontrivial multipliers.* By the same argument as above, we have the following lemma.

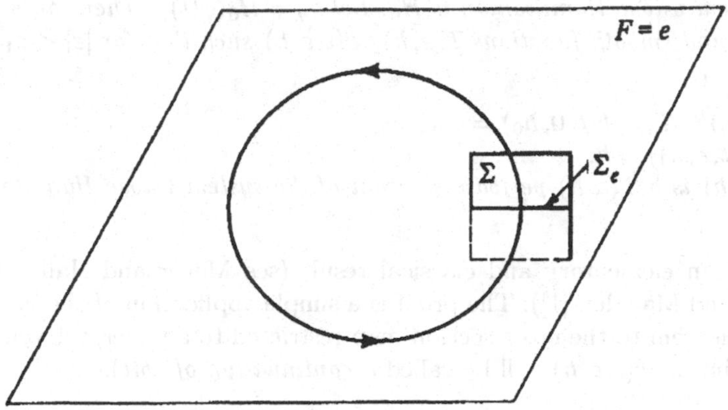

Fig. 6.2. Poincaré map in an integral surface

Lemma 6.5.2. *If the multipliers of the periodic solution of a system with a nondegenerate integral are* $1, 1, \lambda_3, \ldots, \lambda_n$*, then the multipliers of the fixed point in the integral surface are* $\lambda_3, \ldots, \lambda_n$.

Lemma 6.5.3. *If the system is Hamiltonian, then the Poincaré map in an integral surface is symplectic.*

Proof. Use the Hamiltonian flow box theorem (see [51]) to get symplectic flow box coordinates (ξ, η). In this case, $H = \eta_1$ and the equations are $\dot{\xi}_1 = 1, \dot{\xi}_i = 0$ for $i = 2, \ldots, n$, and $\dot{\eta}_i = 0$ for $i = 1, \ldots, n$. The cross section is $\xi_1 = 0$ and the integral parameter is $\eta_1 = e$. The Poincaré map in an integral surface in these coordinates is in terms of the symplectic coordinates $\xi_2, \ldots, \xi_n, \eta_2, \ldots, \eta_n$ on Σ_e. Since the total map $(\xi, \eta) \longrightarrow \phi(T, (\xi, \eta))$ is symplectic, the map $\zeta \longrightarrow Q(e, \zeta)$ is symplectic.

If none of the nontrivial multipliers is 1 and the integral is nondegenerate on the periodic solution, then we say that the periodic solution (or fixed point) is *elementary* or *nondegenerate.*

Theorem 6.5.1. *(Cylinder Theorem) An elementary periodic orbit of a system with an integral lies in a smooth cylinder of periodic solutions parameterized by the integral F. (See Figure 6.3.)*

Proof. Apply the implicit function theorem to $Q(e,y) - y = 0$ to get a one-parameter family of fixed points $y^*(e)$ in each integral surface $F = e$.

In the same manner we have the following important perturbation theorem.

Theorem 6.5.2. *Let $H_\varepsilon : P_\varepsilon \to \mathbb{R}$ be a smooth one-parameter family of Hamiltonians for $|\varepsilon| \leq \varepsilon_0$. Let $\phi(t)$ be a non-degenerate T-periodic solution of the system whose Hamiltonian is H_0. Let $h_0 = H_0(\phi(t))$. Then there exist an $\varepsilon_1 > 0$ and smooth functions $T(\varepsilon, h)$, $\Phi(t, \varepsilon, h)$ such that for $|\varepsilon| < \varepsilon_1$ and $|h - h_0| < \varepsilon_1$:*

1. $T(0, h_0) = T, \quad \Phi(t, 0, h_0) = \phi(t),$
2. $H_\varepsilon(\Phi(t, \varepsilon, h)) = h,$
3. *$\Phi(t, \varepsilon, h)$ is a $T(\varepsilon, h)$-periodic solution of the system whose Hamiltonian is H_ε.*

This is an elementary and classical result (see Meyer and Hall [51] or Abraham and Marsden [1]). The proof is a simple application of the implicit function theorem to the cross section map restricted to an energy level.

The solution $\Phi(t, \varepsilon, h)$ will be called a *continuation of* $\phi(t)$.

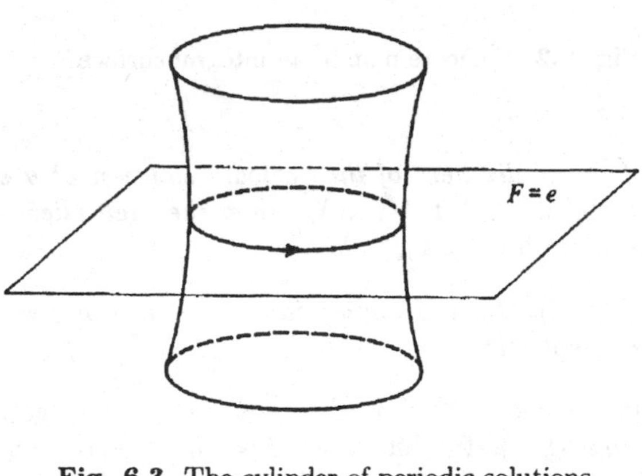

Fig. 6.3. The cylinder of periodic solutions

6.6 Systems with Symmetries

As before, let $\Psi : \mathfrak{G} \times \mathfrak{M} \longrightarrow \mathfrak{M}$ be an action of a Lie group $\mathfrak{G}$ (with algebra $\mathfrak{A}$) on a manifold $\mathfrak{M}$, and let there be a smooth vector field X defined on $\mathfrak{M}$ such that in local coordinates z, X has the form of the differential equation

$$\dot{z} = f(z). \tag{6.11}$$

Let the general solution be $\phi(t, \zeta)$. Assume that $\mathfrak{G}$ is a symmetry group for equation (6.11), so we have

$$\phi(t, \Psi(g, \zeta)) = \Psi(g, \phi(t, \zeta)) \text{ for all } t \in \mathbb{R},\ g \in \mathfrak{G},\ \zeta \in \mathfrak{M}. \tag{6.12}$$

Sometimes it is best to study the flow defined by (6.11) by dropping down to the quotient space, and sometimes it is not. The reduced space may not be a manifold everywhere, and it may be difficult to determine the nature of the reduced space globally. Questions about periodic solutions are local questions, so global assumptions about the reduced space may not be germane to the problem at hand. Here we will look briefly at the periodic solutions on all of $\mathfrak{M}$, not on the reduced space.

The solution $\phi(t, \zeta_0)$ returns to a symmetric configuration at a time T if there is a $g \in \mathfrak{G}$ such that

$$\phi(T, \zeta_0) = \Psi(g, \zeta_0). \tag{6.13}$$

Lemma 6.6.1. *If* (6.13) *holds, then* $\phi(nT, \zeta_0) = \Psi(g^n, \zeta_0)$ *for all* $n \in \mathbb{Z}$.

Proof. The formula holds for $n = 0, 1$. Assume it holds for some $n > 1$. Then

$$\begin{aligned}
\phi((n+1)T, \zeta_0) = \phi(T, \phi(T, \zeta_0)) &= \phi(T, \Psi(g^n, \zeta_0)) \\
= \Psi(g^n, \phi(T, \zeta_0)) &= \Psi(g^n, \Psi(g, \zeta_0)) \\
= \Psi(g^{n+1}, \zeta_0),&
\end{aligned}$$

so the formula holds for positive n. For negative n, note that

$$\phi(T, \Psi(g^{-1}, \zeta_0)) = \Psi(g^{-1}, \phi(T, \zeta_0)) = \Psi(g^{-1}\Psi(g, \zeta_0)) = \zeta_0$$

and

$$\Psi(g^{-1}, \zeta_0)) = \phi(-T, \phi(T, \Psi(g^{-1}, \zeta_0)) = \phi(-T, \zeta_0)$$

and use induction as before.

Thus if a solution returns to a symmetric configuration once, it will do so periodically. There are two cases.

Consider the map $\Psi(\cdot, \zeta_0) : \mathfrak{G} \longrightarrow \mathfrak{M}$. The image of this map, denoted by $\mathfrak{O} \subset \mathfrak{M}$, is the set of images of the point ζ_0 by the group action; it is called the *orbit of* $\mathfrak{G}$ *through* ζ_0. For each $a \in \mathfrak{A}$, $\psi_a(t, \zeta) = \Psi(e^{at}, \zeta)$ is a flow which commutes with $\phi(t, \zeta)$. Let

$$V = \left\{ \frac{d\psi_a(t, \zeta_0)}{dt}\bigg|_{t=0} : a \in \mathfrak{A} \text{ and } ge^{at} \equiv e^{at}g \right\}.$$

V is the set of tangent vectors to $\mathfrak{O}$ at ζ_0. Let $w = \dot{\phi}(0, \zeta_0)$. The two cases are (i) $w \in V$, in which case we call the solution $\phi(t, \zeta_0)$ a *relative equilibrium*, and (ii) $w \notin V$, in which case we call the solution $\phi(t, \zeta_0)$ a *relative periodic solution.*

In order to find a relative periodic solution, one must solve (6.13), or equivalently, solve

$$\Psi(g^{-1}, \phi(T, \zeta))\Big|_{\zeta=\zeta_0} = \zeta_0 \tag{6.14}$$

for g, T, ζ_0. Define the *(characteristic) multipliers* of the relative periodic solution (equilibrium) to be the eigenvalues of

$$\frac{\partial \Psi(g^{-1}, \phi(T, \zeta))}{\partial \zeta}\bigg|_{\zeta=\zeta_0}.$$

Lemma 6.6.2. *Let the dimension of* V *be* s. *Then a relative equilibrium has the multiplier* $+1$ *with multiplicity at least* s *and a relative periodic solution has the multiplier* $+1$ *with multiplicity at least* $s+1$.

Proof. Since all of the actions Ψ, ψ_a, ϕ commute, equation (6.14) implies

$$\Psi(g^{-1}, \phi(T, \psi_a(t, \zeta_0))) \equiv \psi_a(t, \zeta_0).$$

Differentiate this expression with respect to t and set $t = 0$ to get

$$\frac{\partial \Psi(g^{-1}, \phi(T, \zeta_0))}{\partial \zeta} \frac{d\psi_a(t, \zeta)}{dt}\bigg|_{t=0} = \frac{d\psi_a(t, \zeta)}{dt}\bigg|_{t=0}$$

so each vector in V is an eigenvector corresponding to the eigenvalue $+1$. Therefore, the multiplicity is at least s.

We also have

$$\Psi(g^{-1}, \phi(T, \phi(t, \zeta_0))) = \phi(t, \zeta_0).$$

Differentiate this expression with respect to t and set $t = 0$ to get

$$\frac{\partial \Psi(g^{-1}, \phi(T, \zeta_0))}{\partial \zeta} \frac{d\phi(t, \zeta)}{dt}\bigg|_{t=0} = \frac{d\psi(t, \zeta)}{dt}\bigg|_{t=0}$$

so $w = \dot{\phi}(T, \zeta_0)$ is also an eigenvector corresponding to $+1$. If $w \notin V$, then the multiplicity is at least $s+1$.

6.7 Hamiltonian Systems with Symmetries

Assume we are in the same situation as in Section 5.5, that is, let $\mathfrak{M}$ be a symplectic manifold of dimension $2n$, $\Psi : \mathfrak{G} \times \mathfrak{M} \longrightarrow \mathfrak{M}$ a symplectic action of the Lie group $\mathfrak{G}$ of dimension m, $\mathfrak{A}$ the algebra of $\mathfrak{G}$, and $H : \mathfrak{M} \longrightarrow \mathbb{R}$ a Hamiltonian that admits $\mathfrak{G}$ as a symmetry group.

Assume that there are m members of the algebra $\mathfrak{A}$ which giving to m integrals $F_1, \ldots, F_m : \mathfrak{M} \longrightarrow \mathbb{R}$. Let $\mathbf{F} = (F_1, \ldots, F_m)$. Assume that $a \in \mathbb{R}^m$ is a regular value for $\mathbf{F}$, so that $\mathfrak{N} = \mathbf{F}^{-1}(a)$ is a submanifold of $\mathfrak{M}$ of dimension $2n - m$. Let $\mathfrak{G}_a$ be the subgroup of $\mathfrak{G}$ that leaves $\mathfrak{N}$ fixed. Now $\mathfrak{G}_a$ acts on $\mathfrak{N}$ as a symmetry group. Define V and s as in the last section.

Lemma 6.7.1. *Under the assumptions stated above, a relative equilibrium has the multiplier +1 with multiplicity at least $m + s$ and a relative periodic solution has the multiplier +1 with multiplicity at least $m + s + 2$.*

For the spatial N-body problem, we have $m = 3+3 = 6$ and $s = 3+1 = 4$, so a relative periodic solution has the multiplier +1 with multiplicity at least 12. For the planar N-body problem, we have $m = 2+1 = 3$ and $s = 2+1 = 3$, so a relative periodic solution has the multiplier +1 with multiplicity at least 6. These are very degenerate problems!

A sharper, less geometric and more algebraic result is as follows. Let $\mathcal{F}$ be the set of all integrals for the system with Hamiltonian H (so in particular $H \in \mathcal{F}$), $W = \{J\nabla F(\zeta_0) : F \in \mathcal{F}\} \subset \mathbb{R}^{2n}$, $Z = \{u \in W : \{u, W\} = 0\}$.

Theorem 6.7.1. *The geometric multiplicity of the multiplier +1 of a periodic solution is at least $\dim W$. The algebraic multiplicity of the multiplier +1 of a periodic solution is at least $\dim W + \dim Z$.*

Proof. See Meyer 1973 [50]

6.8 Problems

Refer to the problems at the end of Chapter 5 before considering the first few problems here.

1 Consider a system of equations $\dot{u} = f(u)$, $u \in \mathbb{R}^{2n}$ which admits a time reversing symmetry $f(Ru) = -Rf(u)$ where R is an $2n \times 2n$ matrix such that R is similar to $diag\{I_n, -I_n\}$. Let $FIX = \{u \in \mathbb{R}^{2n} : Su = u\}$. Show that FIX is an n dimensional subspace of $\mathbb{R}^{2n}$. Show that if $\phi(t)$ is a solution with $\phi(0) \in FIX$ and $\phi(T) \in FIX$ with $T > 0$, then $\phi(t)$ is a $2T$-periodic solution and the orbit of this periodic solution is carried into itself by R. Such a periodic solution is called a *symmetric periodic solution.*

2 Let S be an anti-symplectic ($S^T JS = -J$), $2n \times 2n$ matrix which is similar to $diag\{I_n, -I_n\}$. Prove that the fixed point set of S, $FIX = \{u \in \mathbb{R}^{2n} : Su = u\}$ is a Lagrangian subspace. A Lagrangian subspace of $\mathbb{R}^{2n}$ is an n-dimensional linear subspace such that the Poisson bracket ($\{u, v\}$) is identically zero.

3 Show that a solution of the restricted problem which crosses the sygyzy axis (the line joining the primaries) perpendicularly at times $t = 0$ and $t = T$, $T > 0$ is a symmetric $2T$ periodic solution.

4 (Gareth Roberts) Show that a solution is a relative equilibrium as defined in the text if and only if it becomes an equilibrium on the reduced space.

7. Satellite Orbits

Here we prove the existence of Poincaré's "periodic orbits of the first kind" by the methods developed in the previous chapters. By first kind, he meant that the solutions were planar and nearly circular. We do not follow Poincaré's original proof exactly. He used a discrete symmetry, arguing that if the three bodies are collinear at time $t = 0$ and again at a time $t = T > 0$, then they will return to their same relative position periodically with period T. That is to say, these solutions are not necessarily periodic in fixed space, but are periodic when the rotational symmetry is eliminated. In fact, Poincaré was proving the existence of periodic solutions on the reduced space, which we call relative periodic solutions. His proof does not give any information about the characteristic multipliers of these solutions and so contains no information about the stability of these orbits. A by-product of the proof given here is that the solutions are elliptic and hence linearly stable.

7.1 Main Problem for Satellite Problem

In celestial mechanics, the "main problem" is the equation of the first approximation, so "defining the main problem" is setting forth the assumptions that yield the correct equations of the first approximation. One of Hill's major contributions to celestial mechanics was redefining the main problem of lunar theory — see Chapter 11 for details. Here we will define the main problem for Poincaré's periodic solutions of the first kind.

Consider a fixed Newtonian frame and let $(\mathbf{q}_0, \mathbf{q}_1, \mathbf{q}_2; \mathbf{p}_0, \mathbf{p}_1, \mathbf{p}_2)$ be the position and momentum vectors, relative to this frame, of three particles having masses m_0, m_1, m_2. In our informal discussions, we shall refer to the particle of mass m_0 as the sun and the particles of masses m_1 and m_2 as the satellites.

Since we wish to eliminate the motion of the center of mass, we choose to represent the equations in Jacobi coordinates $(\mathbf{q}, \mathbf{x}_1, \mathbf{x}_2, \mathbf{G}, \mathbf{y}_1, \mathbf{y}_2)$ and then set $\mathbf{g} = \mathbf{G} = 0$. That is, we perform the following symplectic change of coordinates,

$$\mathbf{x}_1 = \mathbf{q}_1 - \mathbf{q}_0,$$
$$\mathbf{x}_2 = \mathbf{q}_2 - (m_0 + m_1)^{-1}\{m_0\mathbf{q}_0 + m_1\mathbf{q}_1\},$$
$$\mathbf{y}_1 = (m_0 + m_1)^{-1}\{m_0\mathbf{p}_1 - m_1\mathbf{p}_0\},$$
$$\mathbf{y}_2 = (m_0 + m_1 + m_2)^{-1}\{(m_0 + m_1)\mathbf{p}_2 - m_2(\mathbf{p}_0 + \mathbf{p}_1)\},$$

obtaining

$$H = \sum_{i=1}^{2}\left(\frac{\|\mathbf{y}_i\|^2}{2M_i}\right) - \frac{m_0 m_1}{\|\mathbf{x}_1\|} - \frac{m_1 m_2}{\|\mathbf{x}_2 - \alpha_0\mathbf{x}_1\|} - \frac{m_2 m_0}{\|\mathbf{x}_2 + \alpha_1\mathbf{x}_1\|}, \tag{7.1}$$

where

$$M_1 = \frac{m_0 m_1}{(m_0 + m_1)}, \qquad M_2 = \frac{m_2(m_0 + m_1)}{(m_0 + m_1 + m_2)},$$

$$\alpha_0 = \frac{m_0}{(m_0 + m_1)}, \qquad \alpha_1 = \frac{m_1}{(m_0 + m_1)}.$$

The main assumption for this problem is that the sun is much more massive than the satellites, i.e., we scale by

$$m_1 \longrightarrow \varepsilon m_1, \qquad m_2 \longrightarrow \varepsilon m_2$$

so that

$$M_1 = \varepsilon m_1 + O(\varepsilon^2), \qquad M_2 = \varepsilon m_2 + O(\varepsilon^2).$$

Scale the variables by

$$\mathbf{y}_1 \longrightarrow \varepsilon\mathbf{y}_1, \qquad \mathbf{y}_2 \longrightarrow \varepsilon\mathbf{y}_2,$$

which is symplectic with multiplier ε^{-1}, so the Hamiltonian becomes

$$H = \left\{\frac{\|\mathbf{y}_1\|^2}{2m_1} - \frac{m_0 m_1}{\|\mathbf{x}_1\|}\right\} + \left\{\frac{\|\mathbf{y}_2\|^2}{2m_2} - \frac{m_0 m_2}{\|\mathbf{x}_2\|}\right\} + O(\varepsilon). \tag{7.2}$$

To see the above, note that $\|\mathbf{q}_1 - \mathbf{q}_0\| = \|\mathbf{x}_1\|$ and $\|\mathbf{q}_2 - \mathbf{q}_0\| = \|\mathbf{x}_2\| + O(\varepsilon)$. Thus to the first order, the problem is two Kepler problems.

Now change from rectangular coordinates $\mathbf{x}_1, \mathbf{x}_2, \mathbf{y}_1, \mathbf{y}_2$ to polar coordinates $r_1, \theta_1, r_2, \theta_2, R_1, \Theta_1, R_2, \Theta_2$ so that the Hamiltonian becomes

$$H = \frac{1}{2m_1}\left\{R_1^2 + \frac{\Theta_1^2}{r_1^2}\right\} - \frac{m_0 m_1}{r_1} + \frac{1}{2m_2}\left\{R_2^2 + \frac{\Theta_2^2}{r_2^2}\right\} - \frac{m_0 m_2}{r_2} + O(\varepsilon).$$

Total angular momentum $\mathbf{O} = \Theta_1 + \Theta_2$ is an integral and the problem is invariant under rotation, so H is invariant under the transformation $(\theta_1, \theta_2) \longrightarrow (\theta_1 + \gamma, \theta_2 + \gamma)$. Thus to drop down to the reduced space, make the following symplectic change of variables:

$$\phi_1 = \theta_1, \qquad \Phi_1 = \Theta_1 + \Theta_2,$$

$$\phi_2 = \theta_2 - \theta_1, \; \Phi_2 = \Theta_2.$$

Now ϕ_1 is ignorable and Φ_1 is an integral, so on the reduced space ignore ϕ_1 and set $\Phi_1 = c$, a constant. Then local coordinates on the reduced space are $r_1, r_2, \phi_2, R_1, R_2, \Phi_2$ and the Hamiltonian becomes

$$H = \frac{1}{2m_1}\left\{R_1^2 + \frac{(c-\Phi_2)^2}{r_1^2}\right\} - \frac{m_0 m_1}{r_1} + \frac{1}{2m_2}\left\{R_2^2 + \frac{\Phi_2^2}{r_2^2}\right\} - \frac{m_0 m_2}{r_2} + O(\varepsilon).$$

The main problem is the equation obtained from the above Hamiltonian with $\varepsilon = 0$, i.e., the first approximation.

7.2 Continuation of Solutions

Set $\varepsilon = 0$ so that the problem decouples completely and Φ_2 is an integral again. Let $\Phi_2 = b$, $c = a + b$, so the equations of motion on the reduced space when $\varepsilon = 0$ are

$$\dot r_1 = \frac{R_1}{m_1}, \qquad \dot R_1 = \frac{a^2}{m_1 r_1^3} - \frac{m_0 m_1}{r_1^2},$$

$$\dot r_2 = \frac{R_2}{m_1}, \qquad \dot R_2 = \frac{b^2}{m_2 r_2^3} - \frac{m_0 m_2}{r_2^2},$$

$$\dot\phi_2 = \frac{b}{m_2 r_2^2} - \frac{a}{m_1 r_1^2}, \qquad \dot\Phi_2 = 0.$$

The r_i, R_i equations have a critical point at

$$r_1 = \tilde r_1 = \frac{a^2}{m_0 m_1^2}, \qquad R_1 = 0,$$

$$r_2 = \tilde r_2 = \frac{b^2}{m_0 m_2^2}, \qquad R_2 = 0,$$

and the variational equation about this critical point is

$$\ddot r_1 + n_1^2 r_1 = 0, \qquad \ddot r_2 + n_2^2 r_2 = 0,$$

where

$$n_1 = \frac{m_0^2 m_1^3}{a^3}, \qquad n_2 = \frac{m_0^2 m_2^3}{b^3}.$$

These quantities are just the mean anomalies for the circular solutions of the Kepler problem. Recall that the mean anomaly of a circular orbit of the Kepler problem is just its frequency.

The equation for ϕ_2 when $r_1 = \tilde r_1, r_2 = \tilde r_2, R_1 = 0, R_2 = 0$ is

$$\dot\phi_2 = n_2 - n_1 = N.$$

Thus the equations when $\varepsilon = 0$ have a periodic solution $r_1 = r_1, r_2 = r_2, R_1 = 0, R_2 = 0, \Phi_2 = c, \phi_2 = Nt$ and its period is

$$T = \frac{2\pi}{N}$$

and its characteristic multipliers are

$$+1, +1, e^{\pm i n_1 T}, e^{\pm i n_2 T}.$$

This periodic solution is nondegenerate (elliptic, in fact), provided only two multipliers are equal to +1, i.e., provided

$$\frac{n_1}{n_2 - n_1} \notin \mathbb{Z}, \qquad \frac{n_2}{n_2 - n_1} \notin \mathbb{Z}. \tag{7.3}$$

If this condition holds, then the periodic solution is nondegenerate when $\varepsilon = 0$ and so, by Theorem 6.5.2, it can be continued into the full three-body problem on the reduced space. Thus we have

Theorem 7.2.1. *When $\varepsilon = 0$, the system whose Hamiltonian is (7.2) consists of two Kepler problems. Circular periodic solutions of these two Kepler problems such that (7.3) holds can be continued into the three-body problem with two small masses as elliptic relative periodic solutions.*

7.3 Problems

1 Consider the restricted problem (2.7) with μ considered as a small parameter. When $\mu = 0$ you have the Kepler problem in rotating coordinates. Change to polar coordinates and investigate what happens when the circular orbits of the Kepler problem in rotating coordinates are nondegenerate, and hence can be continued into the restricted problem for μ small. These periodic solutions of the restricted problem correspond to the periodic solutions of the first kind of Poincaré.

2 Show that the elliptic solutions of the Kepler problem in rotating coordinates are degenerate (all the multipliers are +1).

3 Try the same type of scaling on the four (or more) body problem. Assume one mass is finite and the others are order ϵ. Show that the circular solutions when $\epsilon = 0$ are degenerate (too many +1s).

4 Show that there are symmetric periodic solutions of the (N+1)-body problem when one particle has finite mass and the others have mass of order ϵ. See Moulton [58].

5 Show that there are periodic solutions of the spatial three-body problem with masses $m_0, \epsilon m_1, \epsilon m_2$ which are continuation of doubly-symmetric circular orbits of the spatial Kepler problem in rotating coordinates. See Soler [84].

8. The Restricted Problem

Previously we introduced the restricted problem whose Hamiltonian is (2.7). Since there are various "restricted problems" we need to be more precise. In general in a restricted problem one or more particles are assumed to have mass equal to zero (the *infinitesimal(s)* or *satellite(s)*) and several of the particles have finite mass (the *primaries*). The k primaries are assumed to follow some known solution of the k–body problem while the infinitesimals move under the gravitational influence of the primaries. But since the infinitesimals have no mass they have no effect on the motion of the primaries or each other.

What we have called the restricted problem is more properly called the circular restricted three–body problem. The three particles have mass $\mu, 1-\mu, 0$, $\mu > 0$ and so it is restricted. The primaries of mass μ and $1-\mu$ move on a circular solution of the two–body problem and hence are "circular". Since it has been studied extensively by Poincaré, Birkhoff and countless others, the adjectives are usually dropped and it becomes known as "the" restricted problem.

In this chapter we will discuss the classical restricted problem in the plane and in space and then we will introduce the circular restricted $(N+1)$–body problem. We will drop the adjective "circular" in this chapter since this is the only case considered. Non-circular restricted problems are discussed in Chapter 12.

To define the general restricted $(N+1)$–body problem take any planar central configuration $(q_1, \ldots, q_N) = (a_1, \ldots, a_N)$ of the N–body problem. This choice is the selection the primaries. So $(q_1, \ldots, q_N, p_1, \ldots, p_N) = (a_1, \ldots, a_N, \omega m_1 a_1, \ldots, \omega m_N a_N)$ is a relative equilibrium, i.e. an equilibrium point in a rotating coordinate system rotating with angular velocity ω. The rotation is about the origin in $\mathbb{R}^2$ for the planar problem or about the z axis in $\mathbb{R}^3$ for the spatial problem. By scaling the size of the central configuration we will assume that $\omega = 1$. Now place a particle of mass zero (the infinitesimal) in the gravitational field created by the primaries. The motion of the infinitesimal is governed by the equations of motion whose Hamiltonian is

$$H_{RN} = \|\eta\|^2/2 - \xi^T J \eta - \sum_{j=1}^{N} \frac{m_j}{\|a_j - \xi\|}. \tag{8.1}$$

In the planar case $\xi, \eta \in \mathbb{R}^2$ and $J = J_2$ and in the spatial case $\xi, \eta \in \mathbb{R}^3$ and $J = J^*$, where

$$J_2 = \begin{pmatrix} 0 & 1 \\ -1 & 0 \end{pmatrix}, \qquad J^* = \begin{pmatrix} 0 & 1 & 0 \\ -1 & 0 & 0 \\ 0 & 0 & 0 \end{pmatrix}.$$

In this chapter, we show that under mild nonresonance assumptions, a nondegenerate periodic solution of the planar or spatial restricted three–body problem can be continued into the full three–body problem and then we generalize this result to the $(N+1)$–body case. This result follows easily from the standard perturbation result for Hamiltonian systems, Theorem 6.5.2, after the Hamiltonian of the problem with one small mass has been correctly scaled. This scaling shows that the restricted problem is indeed the first approximation of the full problem with one small mass.

Also we shall show that some bifurcation results for the restricted problem can be continued into the three or $(N+1)$–body problem.

8.1 Main Problem for the Three-Bodies

In this section we make a series of symplectic changes of variables in the three–body problem which show that the restricted problem is the limit of the reduced problem with one small mass. The reduced problem with one small mass is separable to the first approximation, i.e. the Hamiltonian of the reduced problem to the first approximation is the sum of the Hamiltonian of the restricted problem and the Hamiltonian of the harmonic oscillator.

The three–body problem in the plane is a six degree of freedom problem and a nine degree of freedom problem in space. By placing the center of mass at the origin and setting linear momentum equal to zero, the planar problem reduces to a four degree of freedom problem and the spatial problem to a six degree of freedom problem. This is easily done by using Jacobi coordinates — see Section 3.5. The Hamiltonian of the three–body problem in rotating (about the z-axis) Jacobi coordinates $(x_0, x_1, x_2, y_0, y_1, y_2)$ is

$$\begin{aligned} H = {} & \frac{\| y_0 \|^2}{2M_0} - x_0^T J y_0 + \frac{\| y_1 \|^2}{2M_1} - x_1^T J y_1 - \frac{m_0 m_1}{\| x_1 \|} + \\ & \frac{\| y_2 \|^2}{2M_2} - x_2^T J y_2 - \frac{m_1 m_2}{\| x_2 - \alpha_0 x_1 \|} - \frac{m_2 m_0}{\| x_2 + \alpha_1 x_1 \|} \end{aligned}$$

where

$$M_0 = m_0 + m_1 + m_2, \quad M_1 = \frac{m_0 m_1}{m_0 + m_1}, \quad M_2 = \frac{m_2(m_0 + m_1)}{m_0 + m_1 + m_2}$$

$$\alpha_0 = \frac{m_0}{m_0 + m_1}, \qquad \alpha_1 = \frac{m_1}{m_0 + m_1}.$$

In the planar problem $x_i, y_i \in \mathbb{R}^2$ and $J = J_2$, whereas, in the spatial problem $x_i, y_i \in \mathbb{R}^3$ and $J = J^*$. In these coordinates x_0 is the center of mass, y_0 is total linear momentum, and total angular momentum is

$$O = x_0 \times y_0 + x_1 \times y_1 + x_2 \times y_2.$$

The set $x_0 = y_0 = 0$ is invariant and setting these two coordinates to zero effects the first reduction. Setting $x_0 = y_0 = 0$ reduces the planar problem by two degrees of freedom and the spatial problem by three degrees of freedom.

Assume that one of the particles has small mass by setting $m_2 = \varepsilon^2$ where ε is to be considered as a small parameter. Also set $m_0 = \mu, m_1 = 1 - \mu$ and $\nu = \mu(1-\mu)$, so that

$$M_1 = \nu = \mu(1-\mu), \qquad M_2 = \varepsilon^2/(1+\varepsilon^2) = \varepsilon^2 - \varepsilon^4 + \cdots,$$

$$\alpha_0 = \mu, \qquad \alpha_1 = 1 - \mu.$$

The Hamiltonian becomes

$$H = K + \tilde{H}$$

where

$$K = \frac{1}{2\nu} \parallel y_1 \parallel^2 - x_1^T J y_1 - \frac{\nu}{\parallel x_1 \parallel},$$

and

$$\tilde{H} = \frac{(1+\varepsilon^2)}{2\varepsilon^2} \parallel y_2 \parallel^2 - x_2^T J y_2 - \frac{\varepsilon^2(1-\mu)}{\parallel x_2 - \mu x_1 \parallel} - \frac{\varepsilon^2 \mu}{\parallel x_2 + (1-\mu) x_1 \parallel}.$$

K is the Hamiltonian of the Kepler problem in rotating coordinates. We can simplify K by making the scaling $x_i \to x_i$, $y_i \to \nu y_i$, $K \to \nu^{-1} K$, $\tilde{H} \to \nu^{-1} \tilde{H}$, $\varepsilon^2 \nu^{-1} \to \varepsilon^2$ so that

$$K = \frac{1}{2} \parallel y_1 \parallel^2 - x_1^T J y_1 - \frac{1}{\parallel x_1 \parallel}. \tag{8.2}$$

and

$$\tilde{H} = \frac{(1+\nu\varepsilon^2)}{2\varepsilon^2} \parallel y_2 \parallel^2 - x_2^T J y_2 - \frac{\varepsilon^2(1-\mu)}{\parallel x_2 - \mu x_1 \parallel} - \frac{\varepsilon^2 \mu}{\parallel x_2 + (1-\mu) x_1 \parallel}. \tag{8.3}$$

We consider angular momentum to be nonzero. In the planar problem we can reduce the problem by one more degree of freedom by holding A fixed and ignoring rotations about the origin. One way to reduce the spatial problem by two more degrees is to hold A fixed and eliminate the rotational symmetry about the A axis. Another way to reduce the spatial problem is to note that A_z, the z-component of angular momentum, and $\mathbf{A} = \parallel A \parallel$, the magnitude of angular momentum, are integrals in involution. It is a classical result that one can reduce a system by two degrees of freedom if there are given two independent integrals in involution [92].

Consider the planar case first. In K change from rectangular coordinates x_1, y_1 to polar coordinates so that

$$K = K_2 = \frac{1}{2}\left\{R^2 + \frac{\Theta^2}{r^2}\right\} - \Theta - \frac{1}{r}, \tag{8.4}$$

where r, θ are the usual polar coordinates in the plane, R is radial momentum and Θ is angular momentum. This problem admits K_2 and Θ as integrals in involution.

K_2 has a critical point at

$$r = 1, \quad \theta = 0, \quad R = 0, \quad \Theta = 1.$$

Expand K_2 in a Taylor series about this critical point, ignore the constant term, and make the scaling

$$r - 1 \longrightarrow \varepsilon r, \qquad \theta \longrightarrow \theta,$$

$$R \longrightarrow \varepsilon R, \qquad \Theta - 1 \longrightarrow \varepsilon^2 \Theta,$$

$$K_2 \longrightarrow \varepsilon^{-2} K_2,$$

to get

$$K_2 = \frac{1}{2}\left\{r^2 + R^2\right\} + O(\varepsilon).$$

Now scale $\tilde{H}$ by the above and

$$x_2 = \xi, \quad y_2 = \varepsilon^2 \eta, \quad \tilde{H} \longrightarrow \varepsilon^{-2} \tilde{H}. \tag{8.5}$$

The totality is a symplectic scaling with multiplier ε^{-2} and so the Hamiltonian of the planar three–body problem becomes $H_R + \frac{1}{2}(r^2 + R^2) + O(\varepsilon)$, where H_R is the Hamiltonian of the restricted three–body problem, i.e.

$$H_R = \frac{1}{2} \parallel \eta \parallel^2 - \xi^T J \eta - \frac{(1-\mu)}{\parallel \xi - (\mu, 0) \parallel} - \frac{\mu}{\parallel \xi + (1-\mu, 0) \parallel}. \tag{8.6}$$

To obtain the above expansion recall $x_1 = (r\cos\theta, r\sin\theta) = (1, 0) + O(\varepsilon)$. Thus in the planar case the Hamiltonian of the reduced three–body problem is to the first approximation the sum of the Hamiltonian of the restricted problem and the Hamiltonian of the harmonic oscillator.

Thus symplectic coordinates on the reduced space are ξ, r, η, R. What do the new coordinates mean? The mass of the satellite is ε^2 and y_2 is its momentum, so η is its velocity. ξ is the satellite's position. r and R measure the deviation of the primaries from a circular path.

We take a different tack for the spatial case. In the spatial case $K = K_3$ has a critical point at $x_1 = a = (1,0,0)^T$, $y_1 = b = (0,1,0)^T$ — it corresponds to a circular orbit of the Kepler problem. Expand K_3 in a Taylor series about this point, ignore the constant term and make the scaling

$$x_1 \to a + \varepsilon u, \qquad y_1 \to b + \varepsilon v, \qquad K_3 \to \varepsilon^{-2} K_3 \tag{8.7}$$

to get $K_3 = K^* + O(\varepsilon)$ where

$$K^* = \frac{1}{2}\left(v_1^2 + v_2^2 + v_3^2\right) + u_2 v_1 - u_1 v_2 + \frac{1}{2}\left(-2u_1^2 + u_2^2 + u_3^2\right). \tag{8.8}$$

Again scale $\tilde{H}$ by (8.7) and (8.5). The totality is a symplectic scaling with multiplier ε^{-2} and so the Hamiltonian of the spatial three–body problem becomes $H_R + K^* + O(\varepsilon)$ where K^* is given in (8.8) and H_R is the Hamiltonian of the spatial restricted problem (i.e. (8.6) with $(\mu, 0)$ and $(1-\mu, 0)$ replaced by $(\mu, 0, 0)$ and $(1-\mu, 0, 0)$).

We have already reduced the spatial problem by using the transitional invariance and the conservation of linear momentum, so now we will complete the reduction by using the rotational invariance and the conservation of angular momentum.

Recall that angular momentum in the original unscaled coordinates is $O = x_1 \times y_1 + x_2 \times y_2$ and in the scaled coordinates it becomes

$$O = (a + \varepsilon u) \times (b + \varepsilon v) + \varepsilon^2 \xi \times \eta \tag{8.9}$$

and so holding angular momentum fixed by setting $A = a \times b$ imposes the constraint

$$0 = a \times v + u \times b + O(\varepsilon) = (-u_3, -v_3, v_2 + u_1) + O(\varepsilon). \tag{8.10}$$

Now let us do the reduction when $\varepsilon = 0$ so that the Hamiltonian is $H = H_R + K^*$ and holding angular momentum fixed is equivalent to $u_3 = v_3 = v_2 + u_1 = 0$. Notice that the angular momentum constraint is only on the q, p variables. Make the symplectic change of variables

$$\begin{aligned} r_1 &= u_1 + v_2, & R_1 &= v_1, \\ r_2 &= u_2 + v_1, & R_2 &= v_2, \\ r_3 &= u_3, & R_3 &= v_3, \end{aligned} \tag{8.11}$$

so that

$$K^* = \frac{1}{2}(r_2^2 + R_2^2) + \frac{1}{2}(r_3^2 + R_3^2) + r_1 R_2 - r_1^2. \tag{8.12}$$

Notice that holding angular momentum fixed in these coordinates is equivalent to $r_1 = r_3 = R_3 = 0$, that R_1 is an ignorable coordinate, and r_1 is an integral. Thus passing to the reduced space reduces K^* to

$$K^* = \frac{1}{2}(r_2^2 + R_2^2). \tag{8.13}$$

Thus when $\varepsilon = 0$ the Hamiltonian of the reduced three–body problem becomes

$$H = H_R + \frac{1}{2}(r^2 + R^2), \tag{8.14}$$

which is the sum of the Hamiltonian of the restricted three–body problem and a harmonic oscillator. Here in (8.14) and henceforth we have dropped the subscript 2. The equations and integrals all depend smoothly on ε and so for small ε the Hamiltonian of the reduced three–body problem becomes

$$H = H_R + \frac{1}{2}(r^2 + R^2) + O(\varepsilon). \tag{8.15}$$

We get the same form for the Hamiltonian in both the planar and the spatial problems. We can also introduce action angle variables (I, ι) by

$$r = \sqrt{2I}\cos\iota, \qquad R = \sqrt{2I}\sin\iota,$$

to give

$$H = H_R + I + O(\varepsilon) \tag{8.16}$$

in both cases.

The reduced three–body problem in two or three dimensions with one small mass is approximately the product of the restricted problem and a harmonic oscillator.

8.2 Continuation of Periodic Solutions

A periodic solution of a conservative Hamiltonian system always has the characteristic multiplier +1 with algebraic multiplicity at least 2. If the periodic solution has the characteristic multiplier +1 with algebraic multiplicity exactly equal to 2 then the periodic solution is called *non-degenerate* or sometimes *elementary.* A non-degenerate periodic solution lies in a smooth cylinder of periodic solutions which are parameterized by the Hamiltonian. Moreover, if the Hamiltonian depends smoothly on parameters then the periodic solution persists for small variations of the parameters — see Chapter 6.

Theorem 8.2.1. *A nondegenerate periodic solution of the planar or spatial restricted three–body problem whose period is not a multiple of 2π can be continued into the reduced three–body problem.*

More precisely:

Theorem 8.2.2. *Let $\eta = \phi(t), \xi = \psi(t)$ be a periodic solution with period T of the restricted problem whose Hamiltonian is (8.6). Let its multipliers be $+1, +1, \beta, \beta^{-1}$ in the planar case or $+1, +1, \beta_1, \beta_1^{-1}, \beta_2, \beta_2^{-1}$ in the spatial case. Assume that $T \neq n2\pi$ for all $n \in \mathbb{Z}$ and $\beta \neq +1$ in the planar case or $\beta_1 \neq +1$ and $\beta_2 \neq +1$ in the spatial case. Then the reduced three–body problem, the system with Hamiltonian (8.15), has a periodic solution of the form $\eta = \phi(t) + O(\varepsilon)$, $\xi = \psi(t) + O(\varepsilon)$, $r = O(\varepsilon)$, $R = O(\varepsilon)$ whose period is $T + O(\varepsilon)$. Moreover, its multipliers are $+1, +1$, $\beta + O(\varepsilon), \beta^{-1} + O(\varepsilon)$, $e^{iT} + O(\varepsilon), e^{-iT} + O(\varepsilon)$ in the planar case or $+1, +1$, $\beta_1 + O(\varepsilon), \beta_1^{-1} + O(\varepsilon)$, $\beta_2 + O(\varepsilon), \beta_2^{-1} + O(\varepsilon)$, $e^{iT} + O(\varepsilon), e^{-iT} + O(\varepsilon)$ in the spatial case.*

Proof. When $\varepsilon = 0$ the reduced problem with Hamiltonian (8.15) has the periodic solution $\eta = \phi(t), \xi = \psi(t), r = 0, R = 0$ with period T. Its multipliers are $+1, +1, \beta, \beta^{-1}, e^{iT}, e^{-iT}$ in the planar case or $+1, +1$, β_1, β_1^{-1}, β_2, β_2^{-1}, e^{iT}, e^{-iT} in the spatial case. By the assumption $T \neq n2\pi$ it follows that $e^{\pm iT} \neq +1$ and so this periodic solution is non-degenerate. The classical continuation theorem (Theorem 6.5.2) can be applied to show that this solution can be continued smoothly into the problem with ε small and non-zero.

The planar version of this theorem is due to Hadjidemetriou [30]. There are similar theorems about non-degenerate symmetric periodic solutions – see [48] and the problems. For a different approach, see [42].

There are three classes of non-degenerate periodic solutions of the restricted problem that are obtained by continuation of the circular orbits of the Kepler problem using a small parameter. The small parameter might be μ, the mass ratio parameter, giving the periodic solutions of the first kind of Poincaré [81, 66], a small distance giving Hill's lunar orbits [13, 19, 81], or a large distance giving the comet orbits [48, 56]. All these papers cited except [48] use a symmetry argument, and so do not calculate the multipliers.

However, in Meyer and Hall [51] a unified treatment of all three cases is given and the multipliers are computed and found to be nondegenerate. Thus, there are three corresponding families of periodic solutions of the reduced problem. The corresponding results with independent proofs for the reduced problem are found in [47, 48, 57, 56, 65, 80].

One of the most interesting families of nondegenerate periodic solution of the spatial restricted problem can be found in Belbruno [11]. He regularized double collisions when $\mu = 0$ and showed that some spatial collision orbits are nondegenerate periodic solutions in the regularized coordinates. Thus, they can be continued into the spatial restricted problem as nondegenerate periodic solutions for $\mu \neq 0$. Now these same orbits can be continued into the reduced three–body problem.

8.3 Bifurcations of Periodic Solutions

Many families of periodic solutions of the restricted problem have been studied and numerous bifurcations have been observed. Most of these bifurcations are 'generic one parameter bifurcations' as defined in [46]: also see [51] Chapter VIII. Other bifurcations seem to be generic in either the class of symmetric solutions or generic two-parameter bifurcations. We claim that these bifurcations can be carried over to the reduced three–body problem *mutatis mutandis.* Since there are a multitude of different bifurcations and they are all generalized in a similar manner we shall illustrate only one simple case — the 3-bifurcation of [46] called the phantom kiss in [1]. My son suggested that 3-bifurcations should be called trifurcations.

Let $p(t, h)$ be a smooth family of non-degenerate periodic solutions of the restricted problem parameterized by H_R , i.e. $H_R(p(t, h)) = h$, with period $\tau(h)$. When $h = h_0$ let the periodic solution be $p_0(t)$ with period τ_0, so $p_0(t) = p(t, h_0)$ and $\tau_0 = \tau(h_0)$. We will say that the τ_0-periodic solution $p_0(t)$ of the restricted problem is a *3-bifurcation orbit* if the cross section map $(\psi, \Psi) \longrightarrow (\psi', \Psi')$ in the surface $H_R = h$ for this periodic orbit can be put into the normal form

$$\psi' = \psi + (2\pi k/3) + \alpha(h - h_0) + \beta\Psi^{1/2}\cos(3\psi) + \cdots$$

$$\Psi' = \Psi - 2\beta\Psi^{3/2}\sin(3\psi) + \cdots$$

$$T = \tau_0 + \cdots$$

and $k = 1, 2$, and α and β are non-zero constants. In the above ψ, Ψ are normalized action-angle coordinates in the cross section intersect $H_R = h$, and T is the first return time for the cross section. The periodic solution, $p(t, h)$, corresponds to the point $\Psi = 0$. The multipliers of the periodic solution $p_0(t)$ are $+1, +1, e^{+2k\pi i/3}, e^{-2k\pi i/3}$ (cube roots of unity) so the periodic solution is a nondegenerate elliptic periodic solution. Thus, this family of periodic solutions can be continued into the reduced problem provided τ_0 is not a multiple of 2π by the result of the last subsection.

The above assumptions imply that the periodic solution $p(t, h)$ of the restricted problem undergoes a bifurcation. In particular, there is a one parameter family, $p_3(t, h)$, of hyperbolic periodic solution of period $3\tau_0 + \cdots$ whose limit is $p_0(t)$ as $h \longrightarrow h_0$. See [46, 51] for complete details.

Theorem 8.3.1. *Let $p_0(t)$ be a 3-bifurcation orbit of the restricted problem that is not in resonance with the harmonic oscillator, i.e. assume that $3\tau_0 \neq 2n\pi$, for $n \in \mathbb{Z}$. Let $\tilde{p}(t, h, \varepsilon)$ be the $\tilde{\tau}(h, \varepsilon)$-periodic solution which is the continuation into the reduced problem of the periodic solution $p(t, h)$ for small ε. Thus $\tilde{p}(t, h, \varepsilon) \longrightarrow (p(t, h), 0, 0)$ and $\tilde{\tau}(h, \varepsilon) \longrightarrow \tau(h)$ as $\varepsilon \longrightarrow 0$.*

Then there is a function $\tilde{h}_0(\varepsilon)$ with $\tilde{h}_0(0) = h_0$ such that $\tilde{p}(t, \tilde{h}_0(\varepsilon), \varepsilon)$ has multipliers $+1, +1, e^{+2k\pi i/3}, e^{-2k\pi i/3}, e^{+\tau i} + O(\varepsilon), e^{-\tau i} + O(\varepsilon)$, i.e. exactly one *pair of multipliers are cube roots of unity. Moreover, there is a family of periodic solutions of the reduced problem, $\tilde{p}_3(t,h,\varepsilon)$ with period $3\tilde{\tau}(h,\varepsilon)+\cdots$ such that $\tilde{p}_3(t,h,\varepsilon) \longrightarrow (p_3(t,h),0,0)$ as $\varepsilon \longrightarrow 0$ and $\tilde{p}_3(t,h,\varepsilon) \longrightarrow \tilde{p}(t,\tilde{h}_0(\varepsilon),\varepsilon)$ as $h \longrightarrow \tilde{h}_0(\varepsilon)$. The periodic solutions of the family $\tilde{p}_3(t,h,\varepsilon)$ are hyperbolic-elliptic, i.e. they have two multipliers equal to +1, two multipliers which are of unit modulus, and two multipliers which are real and not equal to ± 1.*

Proof. Since the Hamiltonian of the reduced problem is $H = H_R + \frac{1}{2}(r^2 + R^2) + O(\varepsilon)$ we can compute the cross section map for this periodic solution in the reduced problem for $\varepsilon = 0$. Use as coordinates ψ, Ψ, r, R in this cross section and let $\eta = h - h_0$. The period map is $(\psi,\Psi,r,R) \longrightarrow (\psi',\Psi',r',R')$ where

$$\psi' = \psi'(\psi,\Psi,r,R,\eta,\varepsilon) = \psi + (2\pi k/3) + \alpha\eta + \beta\Psi^{1/2}\cos(3\psi) + \cdots$$

$$\Psi' = \Psi'(\psi,\Psi,r,R,\eta,\varepsilon) = \Psi - 2\beta\Psi^{3/2}\sin(3\psi) + \cdots$$

$$\begin{pmatrix} r' \\ R' \end{pmatrix} = \begin{pmatrix} r'(\psi,\Psi,r,R,\eta,\varepsilon) \\ R'(\psi,\Psi,r,R,\eta,\varepsilon) \end{pmatrix} = B\begin{pmatrix} r \\ R \end{pmatrix} + \cdots$$

where

$$B = \begin{pmatrix} \cos\tau & \sin\tau \\ -\sin\tau & \cos\tau \end{pmatrix}.$$

Since the periodic solution of the restricted problem is non-degenerate it can be continued into the reduced problem and so we may transfer the fixed point to the origin, i.e. $\Psi = r = R = 0$ is fixed.

Since $\alpha \neq 0$ we can solve $\psi'(0,0,0,0,\eta,\varepsilon) = 2\pi k/3$ for η as a function of ε to get $\tilde{\eta}(\varepsilon) = h - \tilde{h}_0(\varepsilon)$. This defines the function $\tilde{h}_0$.

Compute the third iterate of the period map to be

$$(\psi,\Psi,r,R) \longrightarrow (\psi^3,\Phi^3,r^3,R^3),$$

where

$$\psi^3 = \psi + 2\pi k + 3\alpha\eta + 3\beta\Psi^{1/2}\cos(3\psi) + \cdots,$$

$$\Psi^3 = \Psi - 2\beta\Psi^{3/2}\sin(3\psi) + \cdots,$$

$$\begin{pmatrix} r^3 \\ R^3 \end{pmatrix} = B^3\begin{pmatrix} r \\ R \end{pmatrix} + \cdots$$

Since $3\tau \neq 2k\pi$ the matrix $B^3 - E$ is nonsingular, where E is the 2×2 identity matrix. Thus we can solve the equations $r^3 - r = 0$, $R^3 - R = 0$ and substitute the solutions into the equations for $\psi^3 - \psi = 0$, $\Psi^3 - \Psi = 0$.

The origin is always a fixed point; so, Ψ is a common factor in the formula for Ψ^3. Since $\beta \neq 0$, the equation $(\Psi^3 - \Psi)/(-2\beta\Psi^{3/2}) = \sin(3\psi) + \cdots$ can be solved for six functions $\psi_j(\Psi,h) = j\pi/3 + \cdots$, $j = 0,1,\ldots,5$. For even

j, $\cos 3\psi_j = +1 + \cdots$, and for odd j, $\cos 3\psi_j = -1 + \cdots$. Substituting these solutions into the ψ equation gives $(\psi^3 - \psi - 2h\pi)/3 = \alpha\eta \pm \beta\Psi^{1/2} + \cdots$. The equations with a plus sign have a positive solution for Ψ when $\alpha\beta\eta$ is negative, and the equations with the negative sign have a positive solution for Ψ when $\alpha\beta\eta$ is positive. The solutions are of the form $\Psi_j^{1/2} = \mp\alpha\eta/\beta$. Compute the Jacobian along these solutions to be

$$\frac{\partial(\Psi^3, \psi^3)}{\partial(\Psi, \psi)} = \begin{pmatrix} 1 & 0 \\ 0 & 1 \end{pmatrix} + \begin{pmatrix} 0 & \mp 6\beta\Psi_j^{3/2} \\ \pm(3\beta/2)\Psi_j^{1/2} & 0 \end{pmatrix},$$

and so the multipliers are $1 \pm 3\alpha^2\eta^2$, and the periodic points are all hyperbolic-elliptic.

There are many other types of generic bifurcations, e.g. Hamiltonian saddle-node bifurcation, period doubling, k-bifurcations with $k > 3$ etc. as listed in [46, 51]. If such a bifurcation occurs in the restricted problem and the period of the basic periodic orbit is not a multiple of 2π then a similar bifurcation takes place in the reduced problem also. The proofs will be essentially the same as the proof given above.

8.4 Main Problem for $(N+1)$-Bodies

Consider the $(N+1)$–body problem with Hamiltonian H_{N+1} in rotating rectangular coordinates (q, p) where the particles are indexed from 0 to N, and make one mass small by setting $m_0 = \varepsilon^2$. Then Hamiltonian (2.5) becomes

$$H_{N+1} = \|p_0\|^2/2\varepsilon^2 - q_0^T J p_0 - \sum_{j=1}^{N} \frac{\varepsilon^2 m_j}{\|q_j - q_0\|} + H_N, \tag{8.17}$$

where H_N is the Hamiltonian of the N–body problem with particles indexed from 1 to N.

Select the primaries by choosing any planar central configuration of the N–body problem, say $(a_1, \ldots, a_N)$. Let $Z = (q_1, \ldots, q_N; p_1, \ldots, p_N)$ and $Z^* = (a_1, \ldots, a_N; -m_1 J a_1, \ldots, -m_N J a_N)$, so Z^* is a relative equilibrium. (Here we have scaled the central configuration so that the frequency ω is 1.) By Taylor's theorem, we have

$$H_N(Z) = H_N(Z^*) + \frac{1}{2}(Z - Z^*)^T S (Z - Z^*) + O(\|Z - Z^*\|^3), \tag{8.18}$$

where S is the Hessian of H_N at Z^*. In (8.17), make the change of variables

$$q_0 = \xi, \qquad p_0 = \varepsilon^2 \eta, \qquad Z = Z^* - \varepsilon V. \tag{8.19}$$

Now $q_i = a_i + O(\varepsilon)$. This change of variables is symplectic with multiplier ε^2 and thus (8.17) becomes

$$H_{N+1} = \left\{ \|\eta\|^2/2 - \xi^T J\eta - \sum_{j=1}^{N} \frac{m_j}{\|a_j - \xi\|} \right\} + \frac{1}{2}V^T SV + O(\varepsilon). \tag{8.20}$$

Thus to the lowest order in ε, the Hamiltonian of the $(N+1)$–body problem decouples into two Hamiltonians, namely, the Hamiltonian of the restricted $(N+1)$–body problem (8.1), and the Hamiltonian of the linearization of the N–body problem about the relative equilibrium Z^*,

$$H_L = \frac{1}{2}V^T SV. \tag{8.21}$$

Thus when $\varepsilon = 0$, the equations of motion are

$$\dot{\xi} = J\xi + \eta,$$
$$\dot{\eta} = J\eta - \sum_{i}^{N} \frac{m_j(a_j - \xi)}{\|a_j - \xi\|^3}, \tag{8.22}$$

and

$$\dot{V} = JSV. \tag{8.23}$$

For the problem of one small mass, these are the equations of the first approximation. (Remember that J is a generic symbol — for the planar problem it is 2×2 in (8.22) and $4N \times 4N$ in (8.23), but for the spatial problem it is 3×3 in (8.22) and $6N \times 6N$ in (8.23).)

8.5 Reduction

Let $M = \varepsilon^2 + m_1 + \ldots + m_N$ and $V = (u_1, \ldots, u_N, v_1, \ldots, v_N)$, so we have $q_i = a_i - \varepsilon u_i$ and $p_i = -m_i J a_i - \varepsilon v_i$. Since the center of mass of the relative equilibrium is fixed at the origin, we have $\sum_1^N m_i a_i = 0$. Thus the center of mass of the system is

$$C = \{\varepsilon^2 \xi - \varepsilon(m_1 u_1 + \ldots + m_N u_N)\}/M, \tag{8.24}$$

linear momentum is

$$L = \varepsilon^2 \eta - \varepsilon(v_1 + \ldots + u_N), \tag{8.25}$$

and angular momentum is

$$O = \varepsilon^2 \xi^T J\eta - \sum_1^N (a_i - \varepsilon u_i)^T J(m_i J a_i + \varepsilon v_i). \tag{8.26}$$

From (8.24), (8.25), and (8.26, we see that the manifold B_ε of the reduced space depends smoothly on ε.

The defining relations of the reduced manifold when $\varepsilon = 0$ are

$$
\begin{aligned}
m_1 u_1 + \ldots + m_N u_N &= 0, \\
v_1 + \ldots + u_N &= 0, \qquad (8.27) \\
\textstyle\sum_1^N \{-u_i^T J(m_i J a_i) + a_i^T J v_i\} &= 0,
\end{aligned}
$$

which are linear constraints on the N–body problem only, so the reduction applies only to the N–body problem when $\varepsilon = 0$.

8.6 Continuation of Periodic Solutions

Now apply Theorem 6.5.2 to the system on the reduced space whose original Hamiltonian is (8.20) to get

Theorem 8.6.1. *Let $\phi(t)$ be a periodic solution of the planar restricted problem* (8.22) *with period τ and characteristic multipliers $1, 1, \beta, \beta^{-1}$, where $\beta \neq 1$. Let the characteristic exponents of the relative equilibrium be $0, 0, \pm i, \pm i, \pm a_5, \ldots, \pm a_N$, where $a_j\tau \not\equiv 0 \bmod 2\pi i$ for $j = 4, \ldots, N$. Then the τ-periodic solution $\xi = \phi(t)$, $V \equiv 0$ of Equations* (8.22) *and* (8.23) *can be continued into the planar $(N+1)$-body problem on the reduced space as a relative periodic solution. Its multipliers are*

$$1, 1, \beta + O(\varepsilon), \beta^1 + O(\varepsilon), \exp \pm i\tau, \exp \pm a_5\tau, \ldots, \exp \pm a_N\tau.$$

By Theorem 6.5.2, it is enough to show that the periodic solution $\xi = \phi(t)$, $V \equiv 0$ is nondegenerate on the reduced space. By Corollary 4.6.6, passing to the reduced space eliminates $0, 0, \pm i, \pm i$ as characteristic exponents of the relative equilibrium, so the characteristic multipliers of this periodic solution are

$$1, 1, \beta^{\pm 1}, \exp \pm i\tau, \exp \pm a_5\tau, \ldots, \exp \pm a_N\tau.$$

Thus the multiplicity of the characteristic multiplier $+1$ is exactly 2 and Theorem 6.5.2 applies.

8.7 Problems

1 Write the Hamiltonian of the restricted four–body problem where the primaries are at the Lagrange equilateral triangle central configuration. Find the equilibria when the masses are all equal to 1.

2 Scale the restricted $(N+1)$–body problem by $\xi \to \varepsilon^{-2}\xi$, $\eta \to \varepsilon\eta$. So ε small means that the infinitesimal is near infinity. Observe that near infinity the Coriolis force dominates and the next most important force looks like a Kepler problem with both primaries at the origin. See [51].

3 Show that there are nearly circular orbits of very large radius of the restricted $(N+1)$–body problem by using the scaling of Problem 2 and Theorem 6.5.2. What does this say about the $(N+1)$–body problem? See Chapter 10.

4 Take the restricted $(N+1)$–body problem and translate one primary to the origin, then scale by $\xi \to \varepsilon^2\xi$, $\eta \to \varepsilon^{-1}\eta$, and $t \to \varepsilon^{-3}t$. So ε small means the infinitesimal is near the primary. Which force is most important, next most important? See [51].

5 Show that there are nearly circular orbits of very small radius of the restricted $(N+1)$–body problem by using the scaling of Problem 4 and Theorem 6.5.2. What does this say about the $(N+1)$–body problem? See Chapter 9.

6 Consider μ as a small parameter in the restricted three–body problem. Show that there are nearly circular orbits for small μ by using Theorem 6.5.2. What does this say about the three–body problem? See [51]. Did you get the same result as in Chapter 7?

7 Consider any one of the generic one-parameter bifurcations in [46] or [51] Chapter VIII. Show that these bifurcations can be continued into the $(N+1)$–body problem.

8 At the Lagrange point $\mathcal{L}_4$ in the restricted three–body problem for $\mu < \mu_1$ there are many bifurcations as discussed in [74, 53]. Show that these bifurcations carry over to the three–body problem. See [54].

9 State and prove the spatial generalization of Theorem 8.6.1. (Hint: Note that the essential fact the needs to found is the generalization of Corollary 4.6.6. In the proof of this Corollary replace polar coordinates by spherical coordinates.)

9. Lunar Orbits

Another method of introducing a small parameter into the $(N+1)$–body problem is to assume that the distance between two of the particles is small. In this case, we shall show that there are periodic solutions in which $N-1$ particles and the center of mass of the other pair move approximately on a relative equilibrium solution, while the pair move approximately on a small circular orbit of the two–body problem about their center of mass.

9.1 Defining the Main Problem

In this Chapter only the planar problem is considered. Consider the $(N+1)$–body problem with Hamiltonian H_{N+1} written in rotating Jacobi coordinates as discussed in Section 3.5. Assume that the center of mass and linear momentum are fixed at the origin, so the Hamiltonian is

$$H_{N+1} = \sum_{i=1}^{N} \left\{ \frac{\|y_i\|^2}{2M_i} - x_i^T J y_i \right\} - \sum_{0 \le i < j \le N} \frac{m_i m_j}{\|d_{ji}\|} \tag{9.1}$$

and total angular momentum is

$$O = \sum_{i=1}^{N} x_i^T J y_i. \tag{9.2}$$

The vector x_1 is the position vector of the first particle relative to the zeroth particle. We wish to consider the case when these two particles are close, so we make the change of variables

$$x_1 = \varepsilon^4 \xi, \tag{9.3}$$

where ε is a small positive parameter. This change of variables is not symplectic, but compensation will be made later. The Hamiltonian becomes

$$H_{N+1} = \frac{\|y_1\|^2}{2M_1} - \varepsilon^4 \xi^T J y_1 - \frac{m_0 m_1}{\varepsilon^4 \|\xi\|} + H_N + O(\varepsilon^4), \tag{9.4}$$

where H_N is the Hamiltonian of the N–body problem in rotating Jacobi coordinates. The N particles have masses $(m_0+m_1), m_2, \dots, m_N$ and the Jacobi coordinates are indexed from 2 to N. Also $M_1 = m_0 m_1/(m_0+m_1)$. Note that $O(\varepsilon^4)$ terms do not contain the momentum terms $y_1, \dots, y_N$. Angular momentum becomes

$$O = \varepsilon^4 x_i^T J y_1 + \sum_{i=2}^{N} x_i^T J y_i. \tag{9.5}$$

The origin of the coordinate system for H_N is the center of mass of the pair with masses m_0, m_1.

Take any planar central configuration $(x_2, \dots, x_N) = (a_2, \dots, a_N)$ of the N–body problem — this selects the approximate path of the center of mass of the pair with masses m_0, m_1 and the $N-1$ particles with masses $m_2, \dots, m_N$. So $(x_2, \dots, x_N, y_2, \dots, y_N) = (a_2, \dots, a_N,\ \omega M_2 a_2, \dots, \omega M_N a_N)$ is a relative equilibrium, i.e. an equilibrium point in a rotating coordinate system rotating with angular velocity ω. By scaling the size of the central configuration we will assume that $\omega = 1$. Define $Z = (x_2, \dots, x_N, y_2, \dots, y_N)$ and let $Z^* = (a_2, \dots, a_N,\ M_2 a_2, \dots, M_N a_N)$ be the corresponding relative equilibrium. Expand H_N in a Taylor series, so

$$H_N(Z) = H_N(Z^*) + \frac{1}{2}(Z - Z^*)^T S (Z - Z^*) + O(\|Z - Z^*\|^3). \tag{9.6}$$

Now change variables by

$$\begin{aligned} \eta &= \varepsilon^{-2} y_1, \\ \varepsilon V &= Z - Z^* \end{aligned} \tag{9.7}$$

and change time and the Hamiltonian by

$$t = \varepsilon^6 \tau, \qquad H_{N+1} - H_N(Z^*) = \varepsilon^{-6} \tilde{H}. \tag{9.8}$$

The composition of (9.3) and (9.7) is a symplectic change of variables with multiplier ε^2, so the new Hamiltonian becomes

$$\tilde{H} = \left\{ \frac{\|\eta\|^2}{2M_1} - \frac{m_0 m_1}{\|\xi\|} \right\} + \varepsilon^6 \left\{ -\xi^T J \eta + \frac{1}{2} V^T S V \right\} + O(\varepsilon^7). \tag{9.9}$$

Thus to the zeroth order in ε, the Hamiltonian $\tilde{H}$ is the Hamiltonian of the Kepler problem,

$$K = \left\{ \frac{\|\eta\|^2}{2M_1} - \frac{m_0 m_1}{\|\xi\|} \right\}$$

and at the sixth order, the rotation term of the Kepler problem and the quadratic terms of the relative equilibrium appear.

The gradient of angular momentum at the relative equilibrium Z^* is nonzero, so the angular momentum integral becomes

$$O = O' + \varepsilon O_1 V + O(\varepsilon^2), \tag{9.10}$$

where O' is $O(Z_0^*)$ and O_1 is the gradient of the angular momentum at Z^* written as a row vector. Holding O fixed is equivalent to holding $\varepsilon^{-1}(O - O') = O_1 V + O(\varepsilon)$ fixed. Thus the reduction to the full reduced space is smooth in ε.

For the moment neglect the $O(\varepsilon^7)$ terms in (9.9) and consider the approximate equations

$$\xi' = \frac{\eta}{M_1} + \varepsilon^6 J\xi, \tag{9.11}$$

$$\eta' = -\frac{m_0 m_1 \xi}{\|\xi\|^3} + \varepsilon^6 J\eta, \tag{9.12}$$

$$V' = J_{2N} S V, \tag{9.13}$$

where $' = d/d\tau$. For the lunar problem, these are the equations of the first approximation.

9.2 Continuation of Periodic Solution

A periodic solution of equations (9.11)-(9.13) is

$$\begin{aligned} \xi^* &= e^{\omega J\tau} a, \\ \eta^* &= M_1 \delta J e^{\omega J\tau} a, \\ V^* &\equiv 0, \end{aligned} \tag{9.14}$$

where $\omega = \delta + \varepsilon^6$, $\delta = \sqrt{m_0 + m_1}$, and a is any constant vector with $\|a\| = 1$. The period map in an energy level is the identity map up to terms of order $O(\varepsilon^5)$, so care must be taken in calculating the characteristic multipliers. Make a periodic change of variables by

$$\xi = e^{\omega J\tau} \zeta, \tag{9.15}$$

so that the first two equations in (9.11)-(9.13) become

$$\zeta'' + 2\delta J\zeta' - \delta^2 \zeta = -\frac{\delta^2 \zeta}{\|\zeta\|^3}. \tag{9.16}$$

The Jacobian of $\zeta/\|\zeta\|^3$ at $a = (1, 0)$ is $R = \begin{pmatrix} -2 & 0 \\ 0 & 1 \end{pmatrix}$, so the linearization of (9.16) about a is

$$\zeta'' + 2\delta J\zeta' - \delta^2 \zeta = -\delta^2 R\zeta, \tag{9.17}$$

from which it is easy to calculate the characteristic polynomial

$$\lambda^2\{\lambda^2 + \delta^2\}. \tag{9.18}$$

Let the relative equilibrium have characteristic exponents

$$0, 0, \pm i, \pm i, \pm i, \pm a_5, \dots, \pm a_{2N},$$

where $a_j \neq 0$ for $j = 5, \dots, N$. In this case, we say that the relative equilibrium is *nondegenerate* or *elementary*. We have already made the first reduction by setting linear momentum and the center of mass equal to zero, so the eigenvalues of $J_{2N}S$ in 9.13 has eigenvalues $0, 0, \pm i, \pm a_5, \dots, \pm a_{2N}$. Then the characteristic exponents of the solutions (9.15) of equations (9.11)-(9.13) are

$$\begin{gathered}1, 1, \exp\left(\pm\frac{i2\pi\delta}{\delta + \varepsilon^6}\right) = 1 \pm \varepsilon^6\frac{2\pi i}{\delta} + O(\varepsilon^{12}),\\ 1, 1, \exp\pm\frac{\varepsilon^6 2\pi i}{\omega},\\ \exp\pm\frac{\varepsilon^6 2\pi a_5}{\omega}, \dots, \exp\pm\frac{\varepsilon^6 2\pi a_{2N}}{\omega}.\end{gathered} \tag{9.19}$$

On the full reduced space, the characteristic multipliers are

$$\begin{gathered}1, 1, 1 \pm \varepsilon^6\frac{2\pi i}{\delta}, 1 \pm \varepsilon^6\frac{2\pi i}{\delta}, 1 \pm \varepsilon^6\frac{2\pi i}{\delta},\\ 1 \pm \varepsilon^6\frac{a_5 2\pi}{\omega}, \dots, 1 \pm \varepsilon^6\frac{a_{2N}2\pi}{\omega},\end{gathered} \tag{9.20}$$

plus items of order ε^{12} or higher. Thus the characteristic multipliers are of the form $1, 1, 1 \pm \varepsilon^6\beta_5 + O(\varepsilon^{12}), \dots, 1 \pm \varepsilon^6\beta_{2N} + O(\varepsilon^{12})$, where $\beta_j \neq 0$ for $j = 5, \dots, 2N$.

In order to continue this solution into the full $(N+1)$–body problem, we must prove an extension of the classical perturbation theorem, Theorem 6.5.2. This extension is very similar to the continuation theorem given in Henrard [33].

Lemma 9.2.1. *Let $\phi_0(t, \varepsilon)$ be a $T_0(\varepsilon)$-periodic solution of a Hamiltonian system with smooth Hamiltonian $L_0(u, \varepsilon)$, where $u \in \mathcal{O}$ is an open set in $\mathbb{R}^{2m}$ and $|\varepsilon| \leq \varepsilon_0$ with characteristic multipliers*

$$1, 1, 1 \pm \varepsilon^p\gamma_2 + O(\varepsilon^{p+1}), \dots, 1 \pm \varepsilon^p\gamma_m + O(\varepsilon^{p+1}),$$

where $\gamma_j \neq 0$ for $j = 2, \dots, m$. Let the period map in an energy level be the identity map up to order ε^{p-1}. Then for any smooth function $\tilde{L}(u, \varepsilon)$, there exist an $\varepsilon_1 > 0$ and smooth functions $T_1(\varepsilon), \phi_1(t, \varepsilon)$ for $|\varepsilon| \leq 1$ such that $\phi_1(t, \varepsilon)$ is a $T_1(\varepsilon)$-periodic solution of the system whose Hamiltonian is $L_1(u, \varepsilon) = L_0(u, \varepsilon) + \varepsilon^{p+1}\tilde{L}(u, \varepsilon)$, where $T_1(\varepsilon) = T_0(\varepsilon) + O(\varepsilon^{p+1})$ and $\phi_1(0, \varepsilon) = \phi_0(0, \varepsilon) + O(\varepsilon^{p+1})$.

Proof. At $\phi_0(0,0) \in \mathcal{O}$, choose a hyperplane transversal to $\phi_0(0,0)$. This hyperplane will be transversal to both flows for ε small enough. Consider the intersections $\sigma_0(\varepsilon)$ and $\sigma_1(\varepsilon)$ of this hyperplane and the level surfaces $L_0(u,\varepsilon) = L_0(\phi_0(0,0),0)$ and $L_1(u,\varepsilon) = L_0(\phi_0(0,0),0)$. For ε small and near $\phi_0(0,0)$, both σ_0 and σ_1 are symplectic manifolds of dimension $2m-2$ and the period maps P_0 and P_1 are defined. Let v be the local coordinates for σ_0 and σ_1 with $v = 0$ corresponding to $\phi_0(0,0)$. The hypothesis gives $P_1 = P_0 + O(\varepsilon^{p+1})$ and $P_0(v,\varepsilon) = v + \varepsilon^p Q(v) + O(\varepsilon^{p+1})$, where $Q(0) = 0$ and the Jacobian matrix of Q at 0 has eigenvalues $\pm\gamma_2, \ldots, \pm\gamma_m$, $\gamma_j \neq 0$. To find a periodic solution of the system with Hamiltonian L_1, one must solve

$$P_1(v,\varepsilon) = v$$

or

$$v + \varepsilon^p Q(v) + O(\varepsilon^{p+1}) = v$$

or

$$Q(v) + O(\varepsilon) = 0.$$

The implicit function theorem implies that this last equation has a smooth solution $\overline{v}(\varepsilon)$ such that $\overline{v}(0) = 0$. The solution $\phi_1(t,\varepsilon)$ is then the solution of the system with Hamiltonian L_1 with initial condition $\overline{v}(\varepsilon)$ at $t = 0$.

This elementary perturbation lemma proves that the solutions (9.15) can be continued into the full $(N+1)$–body problem.

Theorem 9.2.1. *Let Z^* be a nondegenerate relative equilibrium of the N–body problem. Then there are relative periodic solutions of the $(N+1)$–body problem where $N-1$ particles and the center of mass of a binary pair move approximately on the relative equilibrium solution and two particles move approximately on a circular orbit about their center of mass.*

The condition that the relative equilibrium be nondegenerate is very weak. For $N = 2$ or 3, all the relative equilibria are nondegenerate: also, Pacella [62] proved that the collinear relative equilibrium is nondegenerate for all N and all masses. Palmore [63] also has established that almost all central configurations are nondegenerate.

For $N = 2$, the above result gives the so-called Hill solutions of the three–body problem established, by Moulton [57] and also discussed by Siegel [80] and Conley [19]. If the relative equilibrium is the triangular configuration given by Lagrange, then the above establishes the existence of the periodic solutions of the four–body problem given in Crandall [21]. If the relative equilibrium is the collinear configuration of the N–body problem, then the above establishes the existence of the periodic solutions of the $(N+1)$–body problem given in Perron [65].

9.3 Problems

1 Consider the restricted problem (2.7). Translate one primary to the origin, then scale by $\xi \to \varepsilon^2\xi$, $\eta \to \varepsilon^{-1}\eta$, and $t \to \varepsilon^{-3}t$. So ε small means the infinitesimal is near the primary. Which force is most important, next most important? See [51].

2 Show that there are nearly circular orbits of very small radius of the restricted three–body problem by using the scaling of Problem 1 and Theorem 6.5.2. What does this say about the three–body problem? See Chapter 8.

3 Consider Hill's lunar equation (2.9), scale by $\xi \to \varepsilon^2\xi$, $\eta \to \varepsilon^{-1}\eta$, and $t \to \varepsilon^{-3}t$. So ε small means the infinitesimal is near the primary. Which force is most important, next most important? Show that there are nearly circular orbits of very small radius of the Hill's lunar equations. What does this say about the three–body problem? See Chapter 11.

4 Show that there are periodic solutions to the restricted three–body problem which are symmetric with respect to the line of sygyzy and are continuations of elliptic orbits of the Kepler problem. See [10, 4, 5, 48].

5 Show that there are periodic solutions to the Hill's lunar problem (2.9) which are symmetric with respect both coordinate axes and are continuations of elliptic orbits of the Kepler problem. See [10, 4, 5, 48].

10. Comet Orbits

The main result of this chapter is the existence of a family of periodic solutions of the planar $(N+1)$-body problem in which one of the particles is at a great distance from the other N particles. This distant particle will be called the *comet*. In this family of periodic solutions, the other N particles, called the *primaries*, move approximately on a nonresonant relative equilibrium solution of the N-body problem. The comet moves approximately on a circular orbit of the Kepler problem about the center of mass of the primary system.

The small parameter used here is a scale parameter whose smallness indicates that the distances between the primaries are small relative to their distance to the comet. The scaling is a symplectic transformation with multiplier. None of the masses is assumed to be small.

For the three-body problem, these solutions correspond to Hill-type periodic solutions since, in a typical Hill-type solution, as discussed in Chapter 9, two particles are close and one is far away. These periodic solutions of the three-body problem were established in Moulton [56], Siegel [80], and Conley [19] (see Chapter 9). For the four-body problem, Crandall [21] established the existence of this family in which the relative equilibrium of the primaries is the equilateral triangular solution of Lagrange.

This family was established in Meyer [48] for the general $(N+1)$-body problem under the additional assumption that the comet has small mass and later in Meyer [45] the small mass assumption was dropped. Analogs of this family in the restricted $(N+1)$-body problem were discussed in Meyer [47].

This chapter uses essentially the same method as the previous chapters. However, this problem has different degeneracy due to the existence of elliptic periodic orbits near a relative equilibrium, and this degeneracy requires some variations in the old arguments.

10.1 Jacobi Coordinates and Scaling

Since the main assumption to be made in this chapter is that the distance of one of the particles, say the $(N+1)$st, to the center of mass of the other N particles is large, it is convenient to use Jacobi coordinates because one of the Jacobi coordinates, x_N, is precisely the vector from the center of mass of N

of the particles to the $(N+1)$st particle. Also, one of the Jacobi coordinates is the center of mass of the whole system, $g = m_0q_0 + m_1q_1 + \cdots + m_Nq_N$, and its conjugate momentum is the total linear momentum of the system, $G = p_0 + p_1 + \cdots + p_N$. The center of mass will be fixed at the origin and total linear momentum will be set to zero by putting $g = G = 0$. Having so fixed the center of mass and the linear momentum, the Hamiltonian of the $(N+1)$-body problem in rotating Jacobi coordinates is

$$H = H_{N+1} = \sum_{j=1}^{N} \left(\frac{\|y_j\|^2}{2M_j} - x_j^T J y_j \right) - U_{N+1}, \tag{10.1}$$

where the M_i are constants depending only on the masses ($M_k = m_k\mu_{k-1}/\mu_k$, $\mu_k = m_0 + m_1 + \cdots + m_k$). See Section 3.5. Write this Hamiltonian as

$$H = H_{N+1} = K + H_N + H^*, \tag{10.2}$$

where

$$K = \frac{\|y_N\|^2}{2M_N} - x_N^T J y_N - \frac{\mu_{N-1}m_N}{\|x_N\|}, \tag{10.3}$$

$$H_N = \sum_{j=1}^{N-1} \left(\frac{\|y_j\|^2}{2M_j} - x_j^T J y_j \right) - U_N, \tag{10.4}$$

$$H^* = m_N \sum_{j=1}^{N} m_j \left\{ \frac{1}{d_{jN}} - \frac{1}{\|x_N\|} \right\}. \tag{10.5}$$

In the above, K is the Hamiltonian of the Kepler problem in rotating coordinates. H_N is the Hamiltonian of the N-body problem (for the first N particles) in rotating Jacobi coordinates. Lastly, H^* is a error term which is small if the distance between the first N particles is small. We need to prepare the terms in the Hamiltonian before scaling to define the main problem.

10.2 Kepler Problem

Change to polar coordinates (r, θ, R, Θ) in (10.3) by

$$x_N = \begin{pmatrix} r\cos\theta \\ r\sin\theta \end{pmatrix}, \qquad y_N = \begin{pmatrix} R\cos\theta - (\Theta/r)\sin\theta \\ R\sin\theta - (\Theta/r)\cos\theta \end{pmatrix} \tag{10.6}$$

so that K becomes

$$K = \frac{1}{2M_N} \left\{ R^2 + \frac{\Theta^2}{r^2} \right\} - \Theta - \frac{\mu_{N-1}m_N}{r} \tag{10.7}$$

and the equations of motion become

$$\dot{r} = \frac{R}{M_N}, \qquad \dot{R} = \frac{\Theta^2}{M_N r^3} - \frac{\mu_{N-1} m_N}{r^2}, \tag{10.8}$$
$$\dot{\theta} = \frac{\Theta}{M_N r^2} - 1, \qquad \dot{\Theta} = 0.$$

These equations have an equilibrium point at $R = 0, \theta = \theta_0$, where θ_0 is arbitrary, $r_0 = (\mu_{N-1} m_N / M_N)^{1/3}$, $\Theta_0 = M_N^{1/3} (\mu_{N-1} m_N)^{2/3}$. The Hamiltonian of the linear variational equations about this equilibrium point is

$$Q = \frac{1}{2} \left\{ \frac{1}{M_N} P^2 + M_N \rho^2 + \frac{\Phi^2}{\Theta_0} - 2\alpha\rho\Phi \right\} \tag{10.9}$$

and the linear variational equations are

$$\dot{\phi} = \Phi/\Theta_0 - \alpha\rho, \qquad \dot{\Phi} = 0, \tag{10.10}$$
$$\dot{\rho} = P/M_N, \qquad \dot{P} = -M_N P + \alpha\Phi,$$

where $\phi = \delta\theta$, $\Phi = \delta\Phi$, $\rho = \delta r$, $P = \delta P$ are the variations and $\alpha = (M_N/\mu_{N-1} m_N)^{1/3}$. The characteristic equation of the linearized equations is $\lambda^2(\lambda^2 + 1)$ and the exponents are $0, 0, +i, -i$.

10.3 Defining the Main Problem

Consider the full Hamiltonian H_{N+1} in (10.2), where K is (10.7), H_N is (10.4), and H^* is (10.5). Scale the variables, time, and the Hamiltonian as follows:

$$\theta = \theta_0 + \varepsilon\phi, \qquad \Theta = \Theta_0 + \varepsilon\Phi,$$
$$r = r_0 + \varepsilon M_N^{-1/2} \rho, \; R = \varepsilon M_N^{1/2} P, \tag{10.11}$$
$$x_j = \varepsilon^4 \xi_j, \qquad y_j = \varepsilon^{-2} \eta_j \qquad \text{for } j = 1, \ldots, N-1,$$
$$H = \varepsilon^6 H_{N+1}, \qquad t' = \varepsilon^{-6} t.$$

This change of variables is symplectic with multiplier ε^{-2}. We shall drop the prime on t in the future. Now ε small means that primaries are close together and the comet is near the circular orbit of the Kepler problem. The Hamiltonian becomes

$$H = \sum_{j=1}^{N-1} \left\{ \frac{\eta_j\|^2}{2M_j} - \varepsilon^6 \xi_j^T J \eta_j \right\} - \sum_{0 \le j < k \le N-1} \frac{m_j m_k}{d_{jk}} + \tag{10.12}$$
$$\frac{\varepsilon^6}{2} \left\{ P^2 + \rho^2 + \frac{\Phi^2}{\Theta_0} - 2\alpha\rho\Phi \right\} + O(\varepsilon^7).$$

In the above, ignore the terms of order ε^7 for the present. To that order, the Hamiltonian decouples into the sum of two terms: the first is the N-body problem in a slowly rotating coordinate system, and the second is the Hamiltonian (10.9) of the linear variational equations (10.10). In this case, the truncated equations of motion are

$$\begin{aligned} \dot{\xi}_j &= \eta_j/M_j - \varepsilon^6 J\xi_j && \text{for } j = 1,\dots,N-1, \\ \dot{\eta}_j &= -\partial U/\partial \xi_j - \varepsilon^6 J\eta_j && \text{for } j = 1,\dots,N-1, \\ \dot{\phi} &= \varepsilon^6\{\Phi/\Theta_0 - \alpha\rho\}, && \dot{\Phi} = 0, \\ \dot{\rho} &= \varepsilon^6 P, && \dot{P} = \varepsilon^6\{-P + \alpha\Phi\}. \end{aligned} \tag{10.13}$$

Let $a = (a_1,\dots,a_{N-1})$ be the nonresonant central configuration for the N-body problem that was discussed in Chapter 4, i.e., assume that

$$-M_j a_j = \partial_j U(a) \qquad \text{for } j = 1,\dots,N-1. \tag{10.14}$$

Define $b = (b_i,\dots,b_{N-1})$ by $b_j = M_j J a_j$. Now a periodic solution of these truncated equations (10.13) is

$$\begin{aligned} \xi_j(t) &= e^{\omega J t} a_j && \text{for } j = 1,\dots,N-1, \\ \eta_j(t) &= e^{\omega J t} b_j && \text{for } j = 1,\dots,N-1, \\ \phi &= \Phi = \rho = P = 0, \end{aligned} \tag{10.15}$$

where $\omega = 1 - \varepsilon^6$ and the period is $2\pi/\omega = 2\pi(1 + \varepsilon^6 + \cdots)$.

In order to calculate the multipliers of this periodic solution of the equations (10.13), make the periodic change of variables

$$\begin{aligned} \xi_j(t) &= e^{\omega J t} w_j && \text{for } j = 1,\dots,N-1, \\ \eta_j(t) &= e^{\omega J t} z_j && \text{for } j = 1,\dots,N-1. \end{aligned} \tag{10.16}$$

The first two equations in (10.13) become

$$\begin{aligned} \dot{w}_j &= z_j/M_j - J w_j && \text{for } j = 1,\dots,N-1, \\ \dot{z}_j &= -\partial_j U(w) - J z_j && \text{for } j = 1,\dots,N-1. \end{aligned} \tag{10.17}$$

The periodic solution (10.15) becomes $w_j = a_j$, $z_j = b_j$, $\phi = \Phi = \rho = P = 0$. Equations (10.17) are the equations of the N-body problem in rotating coordinates, so the variational equations about $w_j = a_j$, $z_j = b_j$ give rise to the exponents $0, 0, +i, -i, \lambda_5, \dots, \lambda_{4N-4}$. Thus the characteristic multipliers are

$$+1, +1, \exp(i2\pi/\omega), \exp(-i2\pi/\omega),$$
$$\exp(\lambda_5 2\pi/\omega), \ldots, \exp(\lambda_{4N-4} 2\pi/\omega), \tag{10.18}$$
$$+1, +1, \exp(+i\varepsilon^6 2\pi/\omega), \exp(-i\varepsilon^6 2\pi/\omega).$$

The eigenvalues $\lambda_5, \ldots, \lambda_{4N-4}$ are assumed not to be integer multiples of i, so $\exp(\lambda_j 2\pi/\omega) \neq 1$ for small ε for $j = 5, \ldots, 4N-4$. Since $2\pi/\omega = 2\pi(1 + \varepsilon^6 + \cdots)$, it follows that $\exp(\pm i2\pi/\omega) = 1 \pm \varepsilon^6 i2\pi + O(\varepsilon^{12})$ and $\exp(\pm i\varepsilon^6 2\pi/\omega) = 1 \pm \varepsilon^6 i2\pi + O(\varepsilon^{12})$. Thus the multipliers in (10.18) fall into three groups: first there are four equal to $+1$, then four of the form $1 \pm i2\pi\varepsilon^6 + \cdots$, and finally $4N-8$ of the form $\delta_j + O(\varepsilon^6)$, $\delta_j = \exp(2\pi\lambda_j) \neq +1$.

Basically, the argument from here on is straightforward application of classical ideas with one variation. The problem still admits rotational symmetry, so angular momentum is an integral. Passing to the reduced space eliminates two of the multipliers equal to $+1$, leaving two. By considering the cross section map in an energy surface, we eliminate the remaining two, so the implicit function theorem can be applied to find a periodic solution of the $(N+1)$-body problem on the reduced space close to the solution (10.15). The tedium comes from the fact that the multipliers differ from $+1$ at different orders. The remaining discussion treats these difficulties.

10.4 Reduced Space

Hamiltonian (10.2) with K as in (10.7) is invariant under the symplectic symmetry of rotation by τ, i.e.,

$$(x_1, y_1, \ldots, x_{N-1}, y_{N-1}, r, \theta, R, \Theta) \to$$
$$(e^{Jt}x_1, e^{J\tau}y_1, \ldots, e^{J\tau}x_{N-1}, e^{J\tau}y_{N-1}, r, \theta + \tau, R, \Theta), \tag{10.19}$$

and so admits total angular momentum

$$O = \sum_{j=1}^{N-1} x_j^T J y_j + \Theta \tag{10.20}$$

as an integral. In the new scaled variables, angular momentum becomes

$$O = \varepsilon^2 \sum_{j=1}^{N-1} \xi_j^T J \eta_j + \Theta_0 + \varepsilon \Phi. \tag{10.21}$$

By fixing angular momentum equal to Θ_0, one can solve for Φ, to find that

$$\Phi = -\varepsilon \sum_{j=1}^{N-1} \xi_j^T J \eta_j. \tag{10.22}$$

By holding O fixed and ignoring the conjugate angle ϕ, one drops to the reduced space. Symplectic coordinates on the reduced space are

$$\xi_1, \eta_1, \ldots, \xi_N, \eta_N, \rho, P$$

and the Hamiltonian of the N–body problem on the reduced space is

$$H = \sum_{j=1}^{N-1} \left\{ \frac{\|\eta_j\|^2}{2M_j} - \varepsilon^6 \xi_j^T J \eta_j \right\} - \sum_{0 \le j < k \le N-1} \frac{m_j m_k}{d_{jk}} + \frac{\varepsilon^2}{2} \left\{ P^2 + \rho^2 \right\} + O(\varepsilon^7). \tag{10.23}$$

This is essentially the Hamiltonian (10.20) without the terms in ϕ and Φ. To order ε^6, there is a periodic solution:

$$\xi_j(t) = e^{\omega J t} a_j, \qquad \eta(t) = e^{\omega J t} b_j \qquad \text{for } i = 1, \ldots, N-1, \qquad \rho = P = 0. \tag{10.24}$$

As above, we compute the multipliers to be

$$+1, +1, \exp(2\pi i/\omega), \exp(-2\pi i/\omega), \exp(\lambda_5 2\pi/\omega), \ldots, \exp(\lambda_{4N-4} 2\pi/\omega), \exp(+i\varepsilon^6 2\pi/\omega), \exp(-i\varepsilon^6 2\pi/\omega). \tag{10.25}$$

Now the multipliers in (10.25) fall into three groups. First there are two equal to $+1$, then four of the form $1 \pm 2\pi i \varepsilon^6 + \cdots$, and finally $4N-8$ of the form $\delta_j + O(\varepsilon^6)$, $\delta_j = \exp(2\pi\lambda_j) \neq +1$.

10.5 Continuation of Periodic Solution

Consider the scaled Hamiltonian H in (10.23) on the reduced space. Up to order ε^6, the solutions (10.24) are $(2\pi/\omega)$-periodic with multipliers as in (10.25). Consider the Poincaré map Σ in an energy surface with H constant. When we consider the Poincaré map in an energy surface, the last two $+1$ multipliers disappear. The fixed points of Σ correspond to periodic solutions.

First, look at the form of the Poincaré map up to order ε^5. To that order, the period is 2π. Dropping the terms of order ε^6 and higher leaves just the Hamiltonian of the N-body problem in fixed coordinates, since the rotation terms are at order ε^6. In the energy surface through the relative equilibrium, there is a two-dimensional surface filled with the elliptic periodic solutions discussed in Section 4.6. These periodic solutions will all have period 2π also, so the period map fixes points on this two-dimensional surface. Also, up to that order the variables ρ and P are fixed. Thus there is a four-dimensional manifold which is fixed under the period map to order ε^5.

Let σ be a local coordinate in this manifold and τ the complementary coordinate in the energy surface. The point $\sigma = 0, \tau = 0$ corresponds to the fixed point up to order ε^6. Thus the Poincaré map is of the form $\Sigma : (\sigma, \tau) \to (\sigma', \tau')$ with

$$\begin{aligned} \sigma' &= \sigma + \varepsilon^6(E_1\sigma + E_2\tau + S(\sigma, \tau)) + O(\varepsilon^7), \\ \tau' &= A_4\tau + \varepsilon^6(E_3\sigma + E_4\tau + T(\sigma, \tau)) + O(\varepsilon^7), \end{aligned} \tag{10.26}$$

where A_4, E_1, E_2, E_3, E_4 are constant matrices of the appropriate sizes and S and T are smooth functions with $S(0,0) = T(0,0) = 0$. From the discussion of the multipliers, the eigenvalues of A_4 are $\delta_j = \exp(2\pi\lambda_j) \neq +1, j = 5, \ldots, 4N-4$ and the eigenvalues of E_1 are $\pm 2\pi i, \pm 2\pi$i.

To find a fixed point of Σ, we must solve

$$\begin{aligned} 0 &= E_1\sigma + E_2\tau + S(\sigma, \tau) + O(\varepsilon), \\ 0 &= (A_4 - I)\tau + \varepsilon^6(E_3\sigma + E_4\tau + T(\sigma, \tau)) + O(\varepsilon^7). \end{aligned} \tag{10.27}$$

A direct application of the implicit function theorem gives a solution of (10.27) of the form $\sigma = \sigma^*(\varepsilon) = O(\varepsilon^7), \tau = \tau^*(\varepsilon) = O(\varepsilon^7)$, where $\sigma^*$and τ^* are smooth functions of ε for small ε and $\sigma^*(0) = \tau^*(0) = 0$.

Thus we have

Theorem 10.5.1. *Take any nonresonant relative equilibrium solution of the N-body problem. There is a periodic solution of the (N + 1)-body problem on the reduced space in which N of the particles remain close to the relative equilibrium solution and the remaining particle is close to a circular orbit of the Kepler problem encircling the center of mass of the N-particle system.*

Corollary 10.5.1. *There are elliptic periodic solutions on the reduced three-body problem where two of the particles move on nearly circular orbits about their center of mass and the third particle moves on a circular orbit of large radius.*

10.6 Problems

1 Discuss the statement: For the three-body problem the periodic orbits in Chapter 9 and Chapter 10 are the same.

2 Scale the restricted $(N+1)$-body problem by $\xi \to \varepsilon^{-2}\xi$, $\eta \to \varepsilon\eta$. So ε small means that the infinitesimal is near infinity. Near infinity the Coriolis force dominates and the next most important force looks like a Kepler problem with both primaries at the origin. See [51].

3 Show that there are nearly circular orbits of very large radius of the restricted $(N+1)$-body problem by using the scaling of Problem 2 and Theorem 6.5.2. What does this say about the $(N+1)$-body problem? Show that they are elliptic. See Chapter 10.
4 Show that the periodic solutions of Problem 3 are of general twist type and so stable by KAM theory. See [47].
5 Show that there are nearly elliptical symmetric periodic orbits of with arbitrary eccentricity in the restricted three-body problem by using the scaling of Problem 2. See [47, 56].
6 Instead of using the scaling (10.11) try the scaling

$$\theta = \theta_0 + \phi, \qquad \Theta = \Theta_0 + \varepsilon^2 \Phi,$$

$$r = r_0 + \varepsilon M_N^{-1/2} \rho, \; R = \varepsilon M_N^{1/2} P,$$

$$x_j = \varepsilon^4 \xi_j, \qquad y_j = \varepsilon^{-2} \eta_j \qquad \text{for } j = 1, \dots, N-1,$$

$$H = \varepsilon^6 H_{N+1}, \qquad t' = \varepsilon^{-6} t.$$

11. Hill's Lunar Equations

One of Hill's major contributions to celestial mechanics was his reformulation of the main problem of lunar theory: he gave a new definition for the equations of the first approximation for the motion of the moon [34]. Since his equations of the first approximation contained more terms than the older first approximations, the perturbations were smaller and he was able to obtain series representations for the position of the moon that converge more rapidly than the previously obtained series. Indeed, for many years lunar ephemerides were computed from the series developed by Brown, who used the main problem as defined by Hill. Even today, most of the searchers for more accurate series solutions for the motion of the moon use Hill's definition of the main problem.

Before Hill, the main problem consisted of two Kepler problems — one describing the motion of the earth and moon about their center of mass, and the other describing the motion of the sun and the center of mass of the earth-moon system. The coupling terms between the two Kepler problems are neglected at the first approximation. Delaunay used this definition of the main problem for his solution of the lunar problem, but after twenty years of computation was unable to meet the observational accuracy of his time.

In Hill's definition of the main problem, the sun and the center of mass of the earth-moon system still satisfy a Kepler problem, but the motion of the moon is described by a different system of equations known as Hill's lunar equations. Using heuristic arguments about the relative sizes of various physical constants, he concluded that certain other terms were sufficiently large that they should be incorporated into the main problem. This heuristic grouping of terms does not lead to a precise description of the relationship between the equations of the first approximation and the full problem. Even crude error estimates are hard to obtain.

In a popular description of Hill's lunar equations, one is asked to consider the motion of an infinitesimal body (the moon) which is attracted to a body (the earth) fixed at the origin. The infinitesimal body moves in a coordinate system rotating so that the positive x axis points to an infinite body (the sun) infinitely far away. The ratio of the two infinite quantities is taken so that the gravitational attraction of the sun on the moon is finite. The connection between Hill's lunar equations and the full three-body problem is not made clear from this description.

In this chapter, we shall use the method of symplectic scaling of the Hamiltonian to give a precise derivation of the main problem of lunar theory. Under one set of assumptions, we shall derive the main problem used by Delaunay and under another, the main problem as given by Hill. The derivations are precise asymptotic statements about the limiting behavior of the three-body problem and so can be used to give precise estimates on the deviation of the solutions of the first approximation and the full solutions. (The estimates are not sharp in the practical sense.)

These derivations give a mathematically sound justification for the choice of Hill's definition of the main problem. The method of symplectic scaling is the proper method for defining the main problem for any mechanical problem. Using this scaling, we prove that any nondegenerate periodic solution of Hill's lunar equations whose period is not a multiple of 2π can be continued into the full three-body problem on the reduced space.

11.1 Defining the Main Problem

In this section, we shall show how to introduce scaled symplectic coordinates into the three-body problem in such a way that Hill's equations are the equations of the first approximation. We shall explore other scaled variables and see why they lead to poor approximations.

Consider a frame that rotates with constant angular frequency equal to 1 with reference to a fixed Newtonian frame and let $q_0, q_1, q_2; p_0, p_1, p_2$ be the position and momentum vectors relative to the rotating frame of three particles of masses m_0, m_1, m_2. In our informal discussions, we shall refer to the particles of mass m_0, m_1, and m_2 as the earth, moon, and sun, respectively. Since we wish to eliminate the motion of the center of mass and also scale the distance between the earth and moon, we choose to represent the equations in Jacobi coordinates. We shall set linear momentum and the center of mass to zero by letting $g = G = 0$ in Jacobi coordinates. This accomplishes the first reduction. That is, we perform the following symplectic change of coordinates,

$$\begin{aligned}
x_1 &= q_1 - q_0, \\
x_2 &= q_2 - (m_0 + m_1)^{-1}\{m_0 q_0 + m_1 q_1\}, \\
y_1 &= (m_0 + m_1)^{-1}\{m_0 p_1 - m_1 p_0\}, \\
y_2 &= (m_0 + m_1 + m_2)^{-1}\{(m_0 + m_1)p_2 - m_2(p_0 + p_1)\},
\end{aligned}$$

to obtain

$$H = \sum_{i=1}^{2} \left\{ \frac{\|y_i\|^2}{2M_i'} - x_i^T J y_i \right\} - \frac{m_0 m_1}{\|x_1\|} - \frac{m_1 m_2}{\|x_2 - \alpha_0' x_1\|} - \frac{m_0 m_2}{\|x_2 + \alpha_1' x_1\|}, \quad (11.1)$$

where

$$M_1' = (m_0 + m_1)^{-1} m_0 m_1,$$

$$M_2' = (m_0 + m_1 + m_2)^{-1}(m_0 + m_1)m_2,$$

$$\alpha_0' = (m_0 + m_1)^{-1} m_0, \qquad \alpha_1' = (m_0 + m_1)^{-1} m_1.$$

With the Hamiltonian in (11.1) as our starting point, we shall proceed to make various assumptions on the sizes of various quantities until we are led to a definition of the equation of the first approximation for lunar theory. Each of these assumptions leads to a natural scaling of the variables.

The first assumption is that the earth and moon have approximately the same mass, but their masses are small relative to the mass of the sun. To that effect, we let

$$m_0 = \varepsilon^{2c}\mu_0, \qquad m_1 = \varepsilon^{2c}\mu_1, \qquad m_2 = \mu_2, \tag{11.2}$$

where ε is a small positive parameter, μ_0, μ_1, μ_2 are positive constants, and c is a positive integer to be chosen later. Since two of the particles have masses that are of order ε^{2c}, their momenta will be of the same order, provided their velocities are of order 1. Although it is not altogether necessary, it will make the discussion clearer if we scale the momenta first, taking this observation about the orders of magnitude into account. Thus we make the substitutions $y_1 \longrightarrow \varepsilon^{2c} y_1, y_2 \longrightarrow \varepsilon^{2c} y_2$ in (11.1). With this symplectic change of variables with multiplier ε^{2c}, the Hamiltonian becomes

$$H = H_1 + H_2 + O(\varepsilon^{2c}),$$

$$H_1 = \frac{\|y_1\|^2}{2M_1} - x_1^T J y_1 - \frac{\varepsilon^{2c}\mu_0\mu_1}{\|x_1\|}, \tag{11.3}$$

$$H_2 = \frac{\|y_2\|^2}{2M_2} - x_2^T J y_2 - \frac{\mu_1\mu_2}{\|x_2 - \alpha_0 x_1\|} - \frac{\mu_0\mu_2}{\|x_0 + \alpha_1 x_1\|},$$

where

$$M_1 = (\mu_0 + \mu_1)^{-1}\mu_0\mu_1, \qquad M_2 = \mu_0 + \mu_1,$$

$$\alpha_0 = (\mu_0 + \mu_1)^{-1}\mu_0, \qquad \alpha_1 = (\mu_0 + \mu_1)^{-1}\mu_1. \tag{11.4}$$

Note that the $O(\varepsilon^{2c})$ terms depends only on $\|y_1\|$ and $\|y_2\|$.

The next assumption is that the distance between the earth and moon ($\|x_1\| = \|q_1 - q_0\|$) is small relative to the distance between the sun and the center of mass of the earth-moon system ($\|x_2\|$). We effect this assumption by making the change of variables $x_1 \longrightarrow \varepsilon^{2a} x_1$, where a is a positive integer to be chosen later. This is not a symplectic change of variables; it will be corrected with further changes of variables given below. This change of variables makes H_2 in (11.3) independent of x_1 to the lowest order. Specifically, we have

$$H_2 = H_3 + O(\varepsilon^{4a}),$$
$$H_3 = \frac{\|y_2\|^2}{2M_2} - x_2^T J y_2 - \frac{\mu_2(\mu_0 + \mu_1)}{\|x_2\|}. \tag{11.5}$$

Note that the term of order ε^{2a} is zero due to the particular form of the constants α_0 and α_1. H_3 is the Hamiltonian of the Kepler problem, where a fixed body of mass μ_2 is located at the origin and another body of mass $\mu_0 + \mu_1$ moves in a rotating frame and is attracted to the fixed body by Newton's law of gravity. One can think of the fixed body as the sun and the other body as the union of the earth and moon.

The third and final assumption that we shall make is that the center of mass of the earth-moon system moves on a nearly circular orbit about the sun. Thus we need to prepare H_3 before making this assumption by a change of coordinates. Since H_3 is the Hamiltonian of a Kepler problem in rotating coordinates, one of the circular orbits becomes a circle of critical points for H_3. Specifically, H_3 has a critical point $x_2 = g$, $y_2 = -M_2 Jg$ for any constant vector d satisfying $\|d\|^3 = \mu_2$. We introduce coordinates

$$Z = \begin{pmatrix} x_2 \\ y_2 \end{pmatrix}$$

and a constant vector

$$Z_0 = \begin{pmatrix} d \\ -M_2 Jd \end{pmatrix}$$

so that H_3 is a function of Z and $\nabla H_3(Z_0) = 0$. By Taylor's theorem, we have

$$H_3(Z) = H_3(Z_0) + \frac{1}{2}(Z - Z_0)^T S(Z - Z_0) + O(\|Z - Z_0\|^3), \tag{11.6}$$

where S is the Hessian of H_3 evaluated at Z_0. Since constants are lost in the formation of the equations of motion, we shall ignore the constant $H_3(Z_0)$ in our further discussions. Thus since we seek solutions that are nearly circular, we seek solutions where Z is close to Z_0: so we make the change of variables $Z - Z_0 \longrightarrow \varepsilon^b V$, where b is again a positive integer to be chosen later.

So far, starting with (11.3), we have proposed the following changes of variables: $x_1 \longrightarrow \varepsilon^{2a} x_1$ and $Z - Z_0 \longrightarrow \varepsilon^b V$. In order to have a symplectic change of variables (of multiplier ε^{-2b}), we must make the further change $y_1 \longrightarrow \varepsilon^{2(b-a)} y_1$. Therefore, we propose the following symplectic change of variables in (11.3):

$$
\begin{aligned}
x_1 &\longrightarrow \varepsilon^{2a} x_1, \\
y_1 &\longrightarrow \varepsilon^{2(b-a)} y_1, \\
Z - Z_0 &\longrightarrow \varepsilon^{b} V, \\
H &\longrightarrow \varepsilon^{-2b} H.
\end{aligned}
\tag{11.7}
$$

Moreover, we have introduced three positive integers a, b, and c as measures of the order of magnitude of three physical quantities. One of the variables a, b, or c could be fixed, but since we seek integer solutions it is best not to choose one of them too early.

Consider the main problem as defined by Delaunay. In this case, the earth-moon system is a Kepler problem, so we must choose the scaling so that the kinetic energy and potential energy in H_1 are of the same order of magnitude. This leads to the restriction that $2b = a + c$. Also, the difference between H_2 and H_3, which is of order ε^{4a}, must be of higher order than either of the energy terms in H_1. This leads to the inequality $2a > b$.

Since the equality $2b = a + c$ and the inequality $2a > b$ do not lead to a unique solution, we choose a small solution in integers, say, $a = 2$, $b = 3$, $c = 4$. With this choice, the Hamiltonian (11.1) becomes

$$
H = \varepsilon^{-2}\left\{\frac{\|y_1\|^2}{2M_1} - \frac{\mu_0\mu_1}{\|x_1\|}\right\} + \left\{\frac{1}{2}V^T S V - x_1^T J y_1\right\} + O(\varepsilon^2). \tag{11.8}
$$

Other choices of a, b, and c consistent with the two constraints lead to qualitatively similar scaled Hamiltonians: that is, the terms $V^T S V$ and $x_1^T J y_1$ are always of order zero and the terms $\|y_1\|^2$ and $1/\|x_1\|$ are of order ε^{2b-4a}, which has a negative exponent. In order to better understand this transformed Hamiltonian, let us make one further change of variables. Define a new time by $\tau = \varepsilon^{-2} t$ and thus a new Hamiltonian by $K = \varepsilon^2 H$, so that the problem defined in the new time is given by

$$
K = \frac{\|y_1\|^2}{2M_1} - \frac{\mu_0\mu_1}{\|x_1\|} + \varepsilon^2\left\{\frac{1}{2}V^T S V - x_1^T J y_1\right\} + O(\varepsilon^4). \tag{11.9}
$$

From the general theory of ordinary differential equations, neglecting a term of order ε^4 in the worst possible case leads to an error of the form $O(\varepsilon^4)e^{L\tau} = O(\varepsilon^4)e^{Lt/\varepsilon^2}$, where L is a constant, so neglecting the higher order terms is only valid for very short times. Since any choice of a, b, and c consistent with the constraints leads to the same qualitative form for the Hamiltonian, there is no way to overcome this difficulty. Clearly we must drop the inequality $2a > b$ and incorporate more terms into the main problem.

Let us proceed to define the main problem as suggested by Hill. Since we want the two energy terms in H_1 to have the same order of magnitude, we still impose the restriction $2b = a + c$. The essential problem in the previous attempt was the fact that H_3 was not a good enough approximation of H_2. Following Hill, we expand the two troublesome terms in H_2 in a Legendre series as follows:

$$\frac{\mu_1\mu_2}{\|x_2-\alpha_0x_1\|}+\frac{\mu_0\mu_2}{\|x_2+\alpha_1x_1\|}$$

$$=\frac{\mu_2(\mu_0+\mu_1)}{\|x_2\|}+\frac{1}{\|x_2\|}\sum_{k=2}^{\infty}d_k\rho^kP_k(\cos\theta),$$

where $\rho=\|x_1\|/\|x_2\|$, $d_k=\mu_1\mu_2\alpha_0^k+\mu_0\mu_2(-\alpha_1)^k$, θ is the angle between x_1 and x_2, and P_k is the kth Legendre polynomial. See [22]. Thus (11.3) becomes

$$H=H_1+H_2-\frac{1}{\|x_2\|}\sum_{k=2}^{\infty}d_k\rho^kP_k(\cos\theta)+O(\varepsilon^{2c}). \tag{11.10}$$

Hill said that the first term in the series should be of the same order of magnitude as the terms in H_1; this leads to the conditions $2a=b$ and $2b=a+c$. The simplest solution in integers is $a=1$, $b=2$, $c=3$. With this choice of scale factors, the Hamiltonian becomes

$$\begin{aligned}H=\frac{\|y_1\|^2}{2M_1}-x_1^TJy_1-\frac{\mu_0\mu_1}{\|x_1\|}-\frac{d_2}{\mu_2}\|x_1\|^2P_2(\cos\theta)\\+\frac{1}{2}V^TSV+O(\varepsilon^2).\end{aligned} \tag{11.11}$$

(Recall $\|x_2\|^3=\mu_2+\cdots$.) Now from the general theory of differential equations, neglecting the $O(\varepsilon^2)$ terms leads to an error of order ε^2 on a bounded time interval. Thus defining the main problem as the Hamiltonian in (11.11) without the $O(\varepsilon^2)$ terms is a far better choice.

In order to reduce the number of constants in (11.11), we shall make one further scaling of the variables. We shall introduce new variables ξ and η to eliminate the subscripts and use the fact that $P_2(x)=\frac{1}{2}(1-3x^2)$. Also, we choose $d=(\mu_2^{1/3},0)$ so that the abscissa points at the sun. Make the symplectic change of coordinates

$$\begin{aligned}x_1&=(\mu_0+\mu_1)^{1/3}\xi,\\y_1&=(\mu_0+\mu_1)^{1/3}M_1\eta,\\V&=(\mu_0+\mu_1)^{1/3}M_1^{1/2}V\end{aligned} \tag{11.12}$$

so that (11.11) becomes

$$\begin{aligned}H=\frac{\|\eta\|^2}{2}-\xi^TJ\eta-\frac{1}{\|\xi\|}-\frac{1}{2}(3\xi_1^2-\|\xi\|^2)\\+V^TSV+O(\varepsilon^2).\end{aligned} \tag{11.13}$$

Our choice of scaled variables has eliminated all the parameters in Hill's equations. Note that we have fixed the time scale by requiring that the period of the sun's motion be 2π.

11.2 Continuation of Periodic Solution

Hill proposed to construct a lunar theory by first finding a periodic solution of the system defined by the Hamiltonian

$$H_L = \frac{1}{2}\|\eta\|^2 - \xi^T J\eta - \frac{1}{\|\xi\|} - \xi_1^2 + \frac{1}{2}\xi_2^2 \tag{11.14}$$

and then continuing this solution into the full problem. (The equations defined by (11.14) are known as Hill's lunar equations.) We shall justify this procedure by proving

Theorem 11.2.1. *Any nondegenerate periodic solution of Hill's lunar equations whose period is not a multiple of 2π can be continued into the full three-body problem on the reduced space as a relative periodic solution. If the solution of Hill's lunar equations is elliptic (hyperbolic), then its continuation is elliptic (elliptic-hyperbolic) on the reduced space.*

More precisely:

Theorem 11.2.2. *Let $\xi = \phi_0(t)$, $\eta = \psi_0(t)$ be a τ-periodic solution of Hill's lunar equations with characteristic multipliers $1, 1, \beta, \beta^{-1}$. Assume that this solution is nondegenerate, i.e., $\beta \neq 1$ and $\tau \neq n2\pi$ for any integer n. Then there exist smooth functions $\phi(t,\varepsilon) = \phi_0(t) + O(\varepsilon^2)$, $\psi(t,\varepsilon) = \psi(t) + O(\varepsilon^2)$, $\tau(\varepsilon) = \tau + O(\varepsilon^2)$, and $V(\varepsilon) = O(\varepsilon^2)$ yielding a $\tau(\varepsilon)$-periodic relative periodic solution of the three-body problem on the reduced space. Moreover, the characteristic multipliers of this periodic solution on the reduced space are 1, 1, $\exp(\pm i\tau + O(\varepsilon^2)), \beta + O(\varepsilon^2), \beta^{-1} + O(\varepsilon^2)$.*

We have carefully set up the equations so that the proof of this theorem is almost exactly the same as the proof of the analogous theorem for the restricted N-body problem given in Chapter 8, so we shall only outline the proof here.

Proof. The system defined by (11.3) admits the total angular momentum integral

$$O = x_1^T Jy_1 + x_2^T Jy_2. \tag{11.15}$$

As before, let $Z = (x_2, y_2)$ and let d be the row vector that is the gradient of $x_2^T Jy_2$ with respect to Z evaluated at Z_0. Since $Z_0 \neq 0$ it follows that $d \neq 0$. The scaling reduces (11.15) to

$$\begin{aligned} O &= \varepsilon^4 x_1^T Jy_1 + \varepsilon^2 cV + O(\varepsilon^4) \\ &= \varepsilon^2\{dV + O(\varepsilon^2)\}. \end{aligned} \tag{11.16}$$

Thus to lowest order in ε, the angular momentum vector depends only on x_2, and y_2 or the V coordinates: that is, most of the angular momentum is in the sun and earth-moon system. Thus to the lowest order, the elimination of the angular momentum integral and its conjugate variable affects only the x_2, y_2 coordinates.

Introduce polar coordinates in the x_2 plane and extend them to obtain a symplectic coordinate system on the x_2, y_2 space. Call these coordinates r, θ, R, Θ. To lowest order in ε, Θ is the total angular momentum, so when we fix angular momentum and ignore its conjugate variable, we effectively eliminate Θ and θ, reducing (11.13) to

$$H = H_L + \frac{1}{2}\left\{\frac{R^2}{M} + Mr^2\right\} + O(\varepsilon^2) \tag{11.17}$$

(see Chapter 3 or [51]). Thus to zeroth order in ε, the Hamiltonian of the three-body problem decouples into the sum of the Hamiltonian for Hill's lunar problem and the Hamiltonian of a harmonic oscillator.

When $\varepsilon = 0$, the equations of motion defined by (11.17) are decoupled and one easily sees that $\xi = \phi_0(t)$, $\eta = \psi_0(t)$, $R = r = 0$ is a τ-periodic solution with characteristic multipliers 1, 1, β, β^{-1}, $e^{i\tau}$, $e^{-i\tau}$. Since we assume that τ is not a multiple of 2π, this periodic solution has precisely two characteristic multipliers equal to $+1$ and so is nondegenerate. Thus the standard theorem of perturbation analysis, Theorem 6.5.2, says that this solution can be continued as a periodic solution of the full problem on the reduced space when $\varepsilon \neq 0$.

In a similar manner we can consider the spatial version of Hill's lunar equation.

Theorem 11.2.3. *Let $\xi = \phi_0(t)$, $\eta = \psi_0(t)$ be a τ-periodic solution of the spatial Hill's lunar equations with characteristic multipliers* 1, 1, β_1, β_1^{-1}, β_2, β_2^{-1}. *Assume that this solution is nondegenerate, i.e., $\beta_1 \neq 1$, $\beta_2 \neq 1$ and $\tau \neq n2\pi$ for any integer n. Then there exist smooth functions $\phi(t,\varepsilon) = \phi_0(t) + O(\varepsilon^2)$, $\psi(t,\varepsilon) = \psi(t) + O(\varepsilon^2)$, $\tau(\varepsilon) = \tau + O(\varepsilon^2)$, and $V(\varepsilon) = O(\varepsilon^2)$ yielding a $\tau(\varepsilon)$-periodic relative periodic solution of the spatial three-body problem on the reduced space. Moreover, the characteristic multipliers of this periodic solution on the reduced space are* 1, 1, $\exp(\pm i\tau + O(\varepsilon^2))$, $\beta_1 + O(\varepsilon^2)$, $\beta_1^{-1} + O(\varepsilon^2)$, $\beta_2 + O(\varepsilon^2)$, $\beta_2^{-1} + O(\varepsilon^2)$.

11.3 Problems

1 Prove Theorem 11.2.3.

2 Scale the restricted problem to obtain Hill's lunar equation.

a Shift one primary of the restricted problem to the origin by the scaling

$$q_1 \to q_1 + 1 - \mu, \quad q_2 \to q_2, \quad p_1 \to p_1, \quad p_2 \to p_2 + 1 - \mu.$$

b Expand the one term of the potential in a Taylor series to get

$$-\frac{1-\mu}{\|q+(1,0)\|} = -(1-\mu)(1 - x_1 + x_1^2 - \frac{1}{2}x_2^2 + \cdots.$$

c Scale the Hamiltonian of the restricted problem by $q \to \mu^{1/3}, p \to \mu^{1/3}p$ to obtain $H_R = H_L + O(\mu^{1/3})$ where H_R is the Hamiltonian of the restricted problem and H_L is the Hamiltonian of Hill's lunar problem.

3 Using the scaling of Problem 2 show that any nondegenerate periodic solution of Hill's lunar equation can be continued into the restricted problem.

12. The Elliptic Problem

This chapter deals with the planar N-body problem of classical celestial mechanics and its relation to the elliptic restricted problems. This problem, unlike the previous problem, is a periodic Hamiltonian system.

We give a different derivation of the elliptic restricted problem, which gives a restricted problem for each type of solution of the Kepler problem. In particular, any solution of the Kepler problem, be it circular, elliptic, parabolic, or hyperbolic, gives rise to a coordinate system in which the Hamiltonian of the full planar N-body problem is relatively simple. If the solution of the Kepler problem is circular, then this coordinate system is the standard rotating coordinates; if the solution of the Kepler problem is elliptic, this coordinate system is the rotating-pulsating coordinates used in the elliptic restricted three-body problem. The derivation given below stresses the role of the Kepler problem and thus avoids some of the tedious trigonometry of the standard derivation. It is tempting to call these coordinates "Kepler coordinates," but that name has already a well established meaning in celestial mechanics, so these coordinates will be called Apollonius coordinates after Apollonius of Perga (c. 262–200 B.C.), who wrote the definitive book on conic sections. The origins of rotating-pulsating coordinates and the elliptic restricted three-body problem go back to the work of Scheibner [73] and were rediscovered by Nechvile [60] and others. The rotating-pulsating coordinates were used to put the three-body problem in a simple form in Waldvogel [88] for a different goal. The notes in Szebehely [86] have more information on the historical works.

A central configuration of the N-body problem is an equilibrium point in these coordinates, so it will also be called a relative equilibrium. Given any central configuration of the N-body problem and any solution of the Kepler problem, then, there is a restricted $(N+1)$-body problem in which N of the bodies move on the solution of the Kepler problem while maintaining their relative position, which is similar to the central configuration, and an infinitesimal body moves under their gravitational attraction. For example, there is a restricted four-body problem, in which three bodies of arbitrary mass move on hyperbolic orbits of the Kepler problem, so that at each instant they are at the vertices of an equilateral triangle, and a fourth infinitesimal body moves under the gravitational attraction of the other three but does not in turn influence their motions. To my knowledge, the only reference to something other than the circular or elliptic restricted problems is Faintich [25], who considered the hyperbolic restricted three-body problem.

The method of symplectic scaling will be used to give a derivation of such a restricted problem, showing the precise asymptotic relationship between the restricted problem and the full $(N+1)$-body problem. This derivation obviates the proof of the fact that a nondegenerate periodic solution of the elliptic restricted $(N+1)$-body problem can be continued into the full $(N+1)$-body problem under mild nonresonance assumptions. A similar theorem was proved for the circular restricted $(N+1)$-body problem in Chapter 8 and in Meyer [48, 49].

12.1 Apollonius Coordinates

Let us recall some basic formulas from the Kepler problem and its solution. Let $\phi = (\phi_1, \phi_2)$ be any solution of the planar Kepler problem, r the length of ϕ, and c its angular momentum, so that

$$\ddot{\phi} = -\frac{\phi}{\|\phi\|^3}, \qquad r = \sqrt{\phi_1^2 + \phi_2^2}, \qquad c = \phi_1\dot{\phi}_2 - \phi_2\dot{\phi}_1, \tag{12.1}$$

where the independent variable is t, time, and $\dot{} = d/dt, \ddot{} = d^2/dt^2$. Rule out collinear solutions by assuming that $c \neq 0$ and then scale time so that $c = 1$. The units of distance and mass are chosen so that all other constants are 1. In polar coordinates (r, θ), the equations become

$$\ddot{r} - r\dot{\theta}^2 = -1/r^2, \qquad d(r^2\dot{\theta})/dt = dc/dt = r\ddot{\theta} + 2\dot{r}\dot{\theta} = 0. \tag{12.2}$$

Using the fact that $c = r^2\dot{\theta} = 1$ is a constant of motion yields

$$\ddot{r} - 1/r^3 = -1/r^2. \tag{12.3}$$

Equation (12.3) is reduced to a harmonic oscillator $u'' + u = 1$ by letting $u = 1/r$ and changing from time t to τ the true anomaly of the Kepler problem, by $dt = r^2 d\tau$ and $' = d/d\tau$ — see Section 3.7. The general solution is then

$$r = r(\tau) = 1/(1 + e\cos(\tau - \omega)), \tag{12.4}$$

where e and ω are integration constants, e being the eccentricity and ω the argument of the pericenter. When $e = 0$, the orbit is a circle, for $0 < e < 1$, an ellipse, for $e = 1$, a parabola, and for $e > 1$, a hyperbola. There is no harm in assuming that the argument of the pericenter is zero, so henceforth $\omega = 0$.

Define a matrix A by

$$A = \begin{pmatrix} \phi_1 & -\phi_2 \\ \phi_2 & \phi_1 \end{pmatrix}, \tag{12.5}$$

so $A^{-1} = (1/r^2)A^T$ and $A^{-T} = (A^T)^{-1} = (1/r^2)A$, where A^T denotes the transpose of A.

Consider the planar N-body problem in fixed rectangular coordinates $(\mathbf{q}, \mathbf{p})$ given by the Hamiltonian

$$H = H_N = \sum_{i=1}^{N} \frac{\|\mathbf{p}_i\|^2}{2m_i} - U(\mathbf{q}), \quad U(\mathbf{q}) = \sum_{1 \le i < j \le N} \frac{m_i m_j}{\|\mathbf{q}_i - \mathbf{q}_j\|}. \tag{12.6}$$

The vectors $\mathbf{q}_i, \mathbf{p}_i \in R^2$ are the position and momentum of the ith particle with mass $m_i > 0$, where $i = 1, \dots, N$. U is the self-potential.

Apollonius coordinates are the symplectic coordinates defined below by two symplectic coordinate changes. First, make the symplectic change of coordinates

$$\mathbf{q}_i = AX_i, \qquad \mathbf{p}_i = A^{-T}Y_i = (1/r^2)AY_i \qquad \text{for } i = 1, \dots, N. \tag{12.7}$$

Recall that if $H(z)$ is a Hamiltonian and $z = T(t)u$ is a linear symplectic change of coordinates, then the Hamiltonian becomes $H(u) + (1/2)u^T W(t)u$, where W is the symmetric matrix $W = JT^{-1}\dot{T}$. Compute

$$\begin{aligned} W &= \begin{pmatrix} 0 & I \\ -I & 0 \end{pmatrix} \begin{pmatrix} r^{-2}A^T & 0 \\ 0 & A^T \end{pmatrix} \begin{pmatrix} \dot{A} & 0 \\ 0 & (r^{-2}\dot{A} - 2r^{-3}\dot{r}A \end{pmatrix} \\ &= \begin{pmatrix} 0 & -r^{-2}(A^T\dot{A})^T \\ -r^{-2}A^T\dot{A} & 0 \end{pmatrix}. \end{aligned} \tag{12.8}$$

Recall that W is symmetric or use $A^T A = r^2 I$ to get the 1,2 position. Now

$$-r^{-2}A^T\dot{A} = r^{-2}\begin{pmatrix} -r\dot{r} & 1 \\ -1 & -r\dot{r} \end{pmatrix}. \tag{12.9}$$

Note that $\|AX\| = r\|X\|$, so the Hamiltonian becomes

$$H = \frac{1}{r^2}\sum_{i=1}^{N} \frac{\|Y_i\|^2}{2m_i} - \frac{1}{r}U(X) - \frac{\dot{r}}{r}\sum_{i=1}^{N} X_i^T Y_i - \frac{1}{r^2}\sum_{i=1}^{N} X_i^T J Y_i. \tag{12.10}$$

Change the independent variable from time t to τ the true anomaly of the Kepler problem by $dt = r^2 d\tau$, $' = d/d\tau$, $H \to r^2 H$ so that

$$H = \sum_{i=1}^{N} \frac{\|Y_i\|^2}{2m_i} - rU(X) - \frac{r'}{r}\sum_{i=1}^{N} X_i^T Y_i - \sum_{i=1}^{N} X_i^T J Y_i. \tag{12.11}$$

The second symplectic change of variables changes only the momentum by letting

$$X_i = Q_i, \qquad Y_i = P_i + \alpha_i Q_i, \tag{12.12}$$

where the $\alpha_i = \alpha_i(\tau)$ are to be determined. This defines the Apollonius coordinates (Q_i, P_i) for $i = 1, \ldots, N$. To compute the remainder term, consider

$$R_i = \begin{pmatrix} 0 & I \\ -I & 0 \end{pmatrix} \begin{pmatrix} I & 0 \\ -\alpha_i I & I \end{pmatrix} \begin{pmatrix} 0 & 0 \\ \alpha_i' & 0 \end{pmatrix} = \begin{pmatrix} \alpha_i' & 0 \\ 0 & 0 \end{pmatrix}. \tag{12.13}$$

Thus the remainder term is $(1/2)\sum \alpha_i'(\tau) Q_i^T Q_i$ and the Hamiltonian becomes

$$\begin{aligned} H = \sum_{i=1}^{N} \frac{\|P_i\|^2}{2m_i} - rU(Q) + \left(\frac{\alpha_i}{m_i} - \frac{r'}{r}\right) \sum_{i=1}^{N} Q_i^T P_i - \\ \sum_{i=1}^{N} Q_i^T J P_i + \sum_{i=1}^{N} \left(\frac{1}{2}\alpha_i' + \frac{1}{2}\frac{\alpha_i^2}{m_i} - \frac{r'}{r}\alpha_i\right) Q_i^T Q_i. \end{aligned} \tag{12.14}$$

Choose α_i so that the third term on the right in (12.14) vanishes, i.e., take $\alpha_i = m_i r'/r$. To compute the coefficient of $Q_i^T Q_i$ in the last sum in (12.14), note that

$$\left(\frac{r'}{r}\right)' - \left(\frac{r'}{r}\right)^2 = \frac{rr'' - 2r(r')^2}{r^2} = r\frac{d}{d\tau}\left(\frac{r'}{r^2}\right) = r\frac{d\dot{r}}{d\tau} = r^3\ddot{r} = 1 - r \tag{12.15}$$

where the last equality comes from the formula (12.3). Thus the Hamiltonian of the N-body problem in Apollonius coordinates is

$$H = \sum_{i=1}^{N} \frac{\|P_i\|^2}{2m_i} - rU(Q) - \sum_{i=1}^{N} Q_i^T J P_i + \frac{(1-r)}{2} \sum_{i=1}^{N} m_i Q_i^T Q_i, \tag{12.16}$$

and the equations of motion are

$$\begin{aligned} Q'_i &= \frac{P_i}{m_i} - JQ_i, \\ P'_i &= r\frac{\partial U}{\partial Q_i} - JP_i - (1-r) m_i Q_i. \end{aligned} \tag{12.17}$$

These are particularly simple equations considering the complexity of the coordinate change.

12.2 Relative Equilibrium

A *central configuration of the N-body problem* is a solution $(Q_1, \ldots, Q_N)$ of the system of nonlinear algebraic equations

$$\frac{\partial U}{\partial Q_i} + \lambda m_i Q_i = 0 \qquad for\, i = 1, \ldots, N \tag{12.18}$$

for some scalar λ. By scaling the distance, λ may be taken as 1. Thus a central configuration is a geometric configuration of the N particles so that the force on the ith particle is proportional to m_i times the position. This is the usual definition of a central configuration. Define *a relative equilibrium* as a critical point of the Hamiltonian of the N-body problem in Apollonius coordinates. This is slightly different from the usual definition of a relative equilibrium.

Proposition 12.2.1. *The relative equilibria are central configurations.*

Proof. The critical points of (12.16) satisfy

$$\partial H/\partial Q_i = -r\partial U/\partial Q_i + JP_i + (1-r)m_i Q_i = 0,$$

$$\partial H/\partial P_i = P_i/m_i - JQ_i = 0.$$

From the second equation $P_i = m_i J Q_i$. Plugging this into the first equation gives

$$-r\partial U/\partial Q_i - m_i Q_i + (1-r)m_i Q_i = -r\left\{\partial U/\partial Q_i + m_i Q_i\right\} = 0.$$

Since r is positive, this equation is satisfied if and only if $\partial U/\partial Q_i + m_i Q_i = 0$.

12.3 Defining the Main Problem

Consider the $(N+1)$-body problem with particles indexed from 0 to N. Let H_{N+1} and U_{N+1} be the Hamiltonian and self-potential of the $(N+1)$-body problem written in Apollonius coordinates. Consider also the N-body problem with particles indexed from 1 to N with H_N and U_N the Hamiltonian and self-potential of the N-body problem written in Apollonius coordinates. We have

$$H_{N+1} =$$

$$\sum_{i=0}^{N} \frac{\|P_i\|^2}{m_i} - rU_N(Q) - \sum_{i=0}^{N} Q_i^T J P_i + \frac{(1-r)}{2}\sum_{i=0}^{N} m_i Q_i^T Q_i = \tag{12.19}$$

$$\frac{\|P_0\|^2}{2m_0} - r\sum_{j=1}^{N} \frac{m_0 m_j}{\|Q_0 - Q_j\|} - Q_0^T J P_0 + \frac{(1-r)}{2} m_0 Q_0^T Q_0 + H_N.$$

Assume that one mass is small by setting $m_0 = \varepsilon^2$. The zeroth body is known as the *infinitesimal* and the other N bodies are known as the *primaries*. Let

Z be the coordinate vector for the N-body problem, so $Z = (Q_1, \ldots, Q_N, P_1, \ldots, P_N)$, and let $Z^* = (a_1, \ldots, a_N, b_1, \ldots, b_N)$ be any central configuration for the N-body problem. By Proposition 12.2.1, $\nabla H_N(Z^*) = 0$. The Taylor expansion for H_N is

$$H_N(Z) = H_N(Z^*) + \frac{1}{2}(Z - Z^*)^T S(\tau)(Z - Z^*) + \cdots$$

where $S(\tau)$ is the Hessian of H_N at Z^*. Forget the constant term $H(Z^*)$. Change coordinates by

$$Q_0 = \xi, \quad P_0 = \varepsilon^2 \eta, \quad Z - Z^* = \varepsilon V. \tag{12.20}$$

This is a symplectic transformation with multiplier ε^{-2}. Making this change of coordinates in (12.19) yields

$$H_{N+1} = R + \frac{1}{2} V^T S(\tau) V + O(\varepsilon), \tag{12.21}$$

where R is the Hamiltonian of the *conic (i.e., circular, elliptic, etc.) restricted* $(N+1)$*-body problem* given by

$$R = \frac{1}{2}\|\eta\|^2 - r \sum_{i=1}^{N} \frac{m_i}{\|\xi - a_i\|} - \xi^T J \eta + \frac{(1-r)}{2} \xi^T \xi. \tag{12.22}$$

To the zeroth order, the equations of motion are

$$\begin{aligned} \xi' &= \eta + J\xi, \\ \eta' &= -r \sum_{i=1}^{N} \frac{m_i(\xi - a_i)}{\|\xi - a_i\|^3} + J\eta - (1 - r)\xi, \end{aligned} \tag{12.23}$$

$$V' = D(\tau)V, D(\tau) = JS(\tau). \tag{12.24}$$

The equations in (12.23) are the equations of the restricted problem and those in (12.24) are the linearized equations of motion about the relative equilibrium.

When $e = 0$, equations (12.23) and (12.24) are time-independent and (12.22) is the Hamiltonian of the circular restricted N-body problem. In this case, a periodic solution of (12.23) is called *nondegenerate* if exactly two of its multipliers are $+1$. When $0 < e < 1$, equations (12.23) and (12.24) are 2π-periodic in τ and (12.22) is the Hamiltonian of the elliptic restricted N-body problem. In this case, a $2k\pi$-periodic solution of (12.23) is called *nondegenerate* if all four of its multipliers are different from $+1$.

In the classical elliptic restricted three-body problem the masses of the primaries are $m_1 = 1 - \mu > 0, m_2 = \mu > 0$ and they are located at $a_1 = (-\mu, 0), a_2 = (1 - \mu, 0)$. The parameter μ is called the *mass ratio parameter*. Thus the Hamiltonian of the classical elliptic three-body problem is

$$R = \frac{1}{2}\|\eta\|^2 - r\left(\frac{1-\mu}{d_1} + \frac{\mu}{d_2}\right) - \xi^T J\eta + \frac{(1-r)}{2}\xi^T\xi, \tag{12.25}$$

where

$$d_1 = \{(\xi_1+\mu)^2 + \xi_2^2\}^{1/2}, \qquad d_2 = \{(\xi_1 - 1 + \mu)^2 + \xi_2^2\}^{1/2},$$
$$r = r(\tau) = 1/(1 + e\cos\tau), \quad 0 < e < 1. \tag{12.26}$$

12.4 Symmetries and Reduction

Henceforth, we will consider the elliptic case only. For the moment, consider the N-body problem in the original rectilinear coordinates $(\mathbf{q}, \mathbf{p})$ of (12.6). This Hamiltonian is invariant under the symplectic extension of the group of Euclidian motions of the plane. These motions carry a periodic solution to a periodic solution, so periodic solutions are not isolated even in an energy level in which H is constant. A theorem in Chapter 5 states that due to this symmetry the algebraic multiplicity of the multiplier $+1$ of a periodic solution of the N-body problem must be at least 8. Unless these degeneracies are eliminated, the standard methods of perturbation analysis will fail, so we will again drop down to the reduced space. Now turn to the Hamiltonian of the N-body problem in Apollonius coordinates.

Let C be the center of mass, L total linear momentum, and O total angular momentum in Apollonius coordinates, i.e.,

$$C = \sum_1^N m_i Q_i, \quad L = \sum_1^N P_i, \quad O = \sum_1^N Q_i^T J P_i. \tag{12.27}$$

From equations (12.17), it follows that

$$C' = -JC + L, \qquad L' = -(1-r)C - JL, \qquad O' = 0. \tag{12.28}$$

From these equations, we see that C and L satisfy a time-varying linear homogeneous Hamiltonian system of equations, so the set $C = L = 0$ is invariant. From the last equation, angular momentum O is an integral. The Hamiltonian of the N-body problem in Apollonius coordinates, equation (12.16), is still invariant under rotations, so the reduction can be carried out in these coordinates also. That is, the reduction can be accomplished by setting $C = L = 0$, O equal to a nonzero constant and identifying points by $(\mathbf{q}, \mathbf{p}) \sim (\mathbf{q}^\dagger, \mathbf{p}^\dagger)$ where $\mathbf{q}_i = A\mathbf{q}_i^\dagger$, $\mathbf{p}_i = A\mathbf{p}_i^\dagger$, $A \in SO_2$ a rotation matrix.

Let (u, v) be rectangular coordinates in $\mathbb{R}^2 \times \mathbb{R}^2$. If the Hamiltonian $K = (1/2)v^T v$ is written in Apollonius coordinates (C, L), then K becomes $K(C, L) = (1/2)L^T L - C^T JL + ((1-r)/2)C^T C$, which is the Hamiltonian for the first two equations in (12.28). Thus the first two equations in (12.28) are just the equations $\dot{u} = v$, $\dot{v} = 0$ written in Apollonius coordinates, so the characteristic multipliers of this system are all $+1$. Therefore, fixing $C = L = 0$ decreases the multiplicity of the multiplier $+1$ by 4. Holding O fixed and going to the quotient space decreases the multiplicity of the multiplier $+1$ by another 2, by the same argument as given in Chapter 8 or in Meyer [48]. So going to the reduced space decreases the multiplicity of $+1$ by 6.

A relative equilibrium becomes an equilibrium for the Hamiltonian on the reduced space. The nontrivial multipliers of the relative equilibrium are defined in the following way: First consider the linear variational equation about the relative equilibrium on the reduced spaces — this is a linear, 2π-periodic system of dimension $4N-6$. In general, the multiplier $+1$ will have multiplicity 2. The remaining $4N-8$ multipliers will be called the *nontrivial multipliers of the relative equilibrium.*

A solution of the $(N+1)$-body problem is called *reduced periodic of period* T if its projection on the reduced space is periodic of period T. A reduced periodic solution of the $(N+1)$-body problem is called *nondegenerate* if its projection on the reduced space is a periodic solution with multiplier $+1$ of multiplicity 2.

12.5 Continuation of Periodic Solution

There are many theoretical and numeric investigations of periodic solutions in the elliptic three-body problem. See Broucke [14, 15], Moulton [58], Schubart [75, 76], Sergysels-Lamy [78], Shelus [79], Szebehely and Giacaglia [87], and their references. Consider a system of 2π-periodic equations $\xi' = f(\tau, \xi, \varepsilon)$ depending on a parameter ε, and let $\chi(\tau)$ be a $2k\pi$-periodic solution when $\varepsilon = 0$. The solution $\chi(\tau)$ can be *continued* if there is a smooth one-parameter family of $2k\pi$-periodic solutions $\chi^\dagger(\tau, \varepsilon)$ defined for ε small such that $\chi^\dagger(\tau, 0) = \chi(\tau)$.

Theorem 12.5.1. *Let $(\phi(\tau), \psi(\tau))$ be a nondegenerate $2k\pi$-periodic solution of the planar elliptic restricted $(N+1)$-body problem in (12.23) with Hamiltonian (12.22). Let the nontrivial multipliers of the relative equilibrium not be kth roots of unity. Then the $2k\pi$-periodic solution $\xi = \phi(\tau), \eta = \psi(\tau), V = 0$ of (12.23), (12.24) can be continued into the full $(N+1)$-body problem as a nondegenerate reduced periodic solution for small values of $m_0 = \varepsilon^2$.*

Proof. Consider the $(N+1)$-body problem using the notation of Section 12.3. Let $V = (u_1, \ldots, u_N, v_1, \ldots, v_N)$ so that $\mathbf{q}_i = a_i - \varepsilon u_i$, $\mathbf{p}_i = b_i - \varepsilon v_i = -m_i J a_i - \varepsilon v_i$. Since the center of mass of the relative equilibrium is fixed at the origin, we have $\sum_1^N m_i a_i = 0$ and

$$C = \varepsilon^2 \xi + \varepsilon \{ m_1 u \cdots m_N u_N \},$$

$$L = \varepsilon^2 \eta + \varepsilon \{ v_1 + \cdots + u_N \},$$

$$A = \varepsilon^2 \xi^T J \eta + \textstyle\sum_1^N ({}_i - \varepsilon_i^T J (b_i - \varepsilon v_i).$$

From these formulas, it follows that the reduced space depends smoothly on the parameter ε and the Hamiltonian on the reduced space is also smooth in ε.

Remember that the $(N+1)$-body problem is time-independent and a periodic solution can be continued if the eigenvalue $+1$ has multiplicity 2. (This is a simple consequence of the implicit function theorem applied to the Poincaré map in an energy level; see Chapter 6.) By the assumptions above, the $2k\pi$-periodic solution $\xi = \phi(\tau)$, $\eta = \psi(\tau)$, $V = 0$ when $\varepsilon = 0$ has the multiplier $+1$ with multiplicity 2 on the reduced space.

Corollary 12.5.1. *Let $(\phi(\tau), \psi(\tau))$ be a nondegenerate $2k\pi$-periodic solution of the classical, elliptic, restricted three-body problem with Hamiltonian (12.25). Then the $2k\pi$-periodic solution $\xi = \phi(\tau)$, $\eta = \psi(\tau)$, $V = 0$ of (12.23), (12.24) can be continued into the full three-body problem as a nondegenerate reduced periodic solution for small values of $m_0 = \varepsilon^2$.*

Proof. The two-body problem has dimension eight and its reduced space is two-dimensional. Therefore, there are no nontrivial multipliers of the relative equilibrium and so no restriction on them.

12.6 Problems

1 Consider the spatial problem.
   - What is the generalization of Apollonius coordinates for the spatial problems? (Hint: Recall the rotating coordinates in $\mathbb{R}^3$.)
   - Show that Theorem 12.5.1 and Corollary 12.5.1 can be generalized to the spatial problem.

2 Show that it is possible to combine the ideas of this chapter and those of Chapter 11 to define an elliptic Hill's lunar equation. One can show that a nondegenerate periodic $2k\pi$-periodic solution of the elliptic Hill's lunar equation can be continued into the full three-body problem as a nondegenerate periodic solution. See [52].

3 Show that a nondegenerate symmetric $2k\pi$-periodic solution of the classical, elliptic, restricted three-body problem with Hamiltonian (12.25) can be continued into the full three-body problem as a nondegenerate symmetric periodic solution for small values of m_0. See [52].

References

1. R. Abraham and J. Marsden, *Foundations of Mechanics*, Benjamin-Cummings, London, 1978.
2. J. Ángel and C. Simó, Effective stability for periodically perturbed Hamiltonian systems, *NATO Adv. Sci. Inst. Ser. B, Phys. 331*, Plenum, NY, 1994, 243-253.
3. Apollonius of Perga, *On Conic Sections*, (Trans. R. Catesby Taliaferro), University of Chicago Press, Chicago,c. 200 B.C.
4. R. F. Arenstorf, A new method of perturbation theory and its application to the satellite problem of celestial mechanics, *J. Reine Angew. Math.* **221**, 1966, 113-145.
5. ———, New periodic solutions of the plane three-body problem corresponding to elliptic motion in the lunar theory, *J. Diff. Eqs* **4**, 1968, 202-256.
6. ———, Periodic solution of circular-elliptic type in the planar n-body problem, *Celest. Mech.* **17**, 1978, 331-355.
7. V. I. Arnold, *Mathematical Methods of Classical Mechanics*, Springer-Verlag, New York, 1978.
8. V. I. Arnold, V. V. Kozlov, and A. I. Neishtadt, *Encyclopaedia of Mathematical Sciences*, Volume 3, Dynamical Systems III, (Trans. A. Iacob), Springer-Verlag, New York, 1987.
9. D. K. Arrowsmith and C. M. Place, *Ordinary Differential Equations*, Chapman and Hall, London, 1982.
10. R. B. Barrar, Existence of periodic orbits of the second kind in the restricted problem of three bodies, *Astronom. J.* **70(1)**, 1965, 3-4.
11. E. A. Belbruno,A new family of periodic orbits for the restricted problem, *Celestial Mech.*, 25, 1981, 397-415.
12. G. D. Birkhoff, *Dynamical Systems*, Amer. Math. Soc., New York, 1927.
13. ———, The restricted problem of three-bodies, *Rend. Circolo. Mat. Palermo*, 39, 1915, 255-334.
14. R. Broucke, Periodic collision orbits in the elliptic restricted three-body problem, *Celest. Mech.* **3(4)**, 1971, 461-477.
15. ———, Stability of periodic orbits in the elliptic, restricted three-body problem, *AIAA Journal* **7(6)**, 1969, 1003-9.
16. A. Bruno, *Local Methods in Nonlinear Differential Equations*, Springer-Verlag, New York, 1979.
17. ———, *The Restricted 3-Body Problem.* Walter de Gruyer, Berlin, 1994.
18. E. A. Coddington and N. Levinson, *Theory of Ordinary Differential Equations*, McGraw-Hill, New York, 1955.
19. C. Conley, On some new long periodic solutions of the plane restricted three body problem, *Comm. Pure Appl. Math.* **XVI**, 1963, 449-467.
20. C. C. Conley and E. Zehnder, The Birkhoff-Lewis fixed point theorem and a conjecture of V. I. Arnold, *Invent. Math.* **73**,1984, 33-49.

21. M. G. Crandall, Two families of periodic solutions of the plane four-body problem, *Amer. J. Math.* **16**, 1967, 275-318.
22. J. M. A. Danby, *Fundamentals of Celestial Mechanics*, MacMillan Co., New York, 1962.
23. R. W. Easton, Some topology of n-body problem, *J. Diff. Eqs.* **19** (1975), 258-269.
24. ______, Some topology of the three-body problem, *J. Diff. Eqs.* **10** (1971), 371-377.
25. M. Faintich, Applications of the restricted hyperbolic three-body problem to a star-sun-comet system, *Celest. Mech.* **6(1)**, 1972, 22-26.
26. U. Fong and K. R. Meyer, Algebras of integrals, *Revista Colombiana Math.* **IX**, 75-90.
27. J. Franks, Recurrence and fixed point in surface homeomorphisms, *Erg. Theory Dyn. Sys.* **8***, 1988, 99-108.
28. G. Gómez and J. Llibre, A note on a conjecture of Poincaré, Celest. Mech., 24(4), 1981, 335-343.
29. M. C. Gutzwiller and D. S. Schmidt, The motion of the moon as computed by the method of Hill, Brown, and Eckert, *Astronomical Papers — Americal Ephemeris and Nautical Almanac*, XXIII, U.S. Gov. Printing Office, Washington, DC, 1986.
30. J. D. Hadjidemetious, The continuation of periodic orbits from the restricted to the general three-body problem, *Celest. Mech.* **12**, 1975, 155-174.
31. J. K. Hale, *Ordinary Differential Equations*, Wiley-Interscience, New York, 1969.
32. P. Hartman, *Ordinary Differential Equations*, Wiley, New York, 1964.
33. J. Henrard, Lyapunov's center theorem for resonant equilibrium, *J. Diff. Eqs* **14**, 1973, 431-441.
34. G. W. Hill, Researches in the lunar theory, *Amer. J. of Math.* **1**, 1878, 5-26, 129-147, 245-260.
35. M. Hirch, *Differential Topology*, Springer-Verlag, New York, 1976.
36. C. G. J. Jacobi, *Vorlesungen über Dynamik*, Verlag G. Reimer, Berlin, 1884.
37. Keynes MSS 130.6, Book 3; 130.5, Sheet 3 — Newton Ms, in the Keynes collection in the library of king's College, Cambridge, UK.
38. J. L. Lagrange, *Recueil des pièces qui ont remporté les prix de l'Académie de Paris,* **IX**, 1772.
39. J. P. LaSalle and S. Lefschetz, *Stability by Liapunov's Direct Method with Applications*, Academic Press, New York, 1961.
40. A. Laub and K. R. Meyer, Canonical forms for symplectic and Hamiltonian matrices, *Celest. Mech.* **9**, 1974, 312-238.
41. H. I. Levine, *Singularities of differentiable mappings*, Proceedings of Liverpool on Singularities–Symposium I (C. T. C. Wall, Ed.), Lecture Notes in Mathematics No. 192, Springer-Verlag, Berlin/New York, 1971.
42. J. Llibra and D. Saari, Periodic orbits for the planar Newtonian three-body problem coming from the elliptic restricted three-body proble, *Trans. Amer. Math. Soc.* **347**, 1995, 3017-3033.
43. A. Lyapunov, Probleme gńéral de la stabilité du mouvement, *Ann. Math. Studies*, 17, Princeton Univ. Press. Princeton NJ, 1947. (Reproduction of 1892 monograph)
44. J. Marsden and A. Weinstein, Reduction of symplectic manifolds with symmetries, *Rep. Math. Phy.* **5**, 1974, 121-130
45. K. R. Meyer, Comet like periodic orbits in the N-body problem, *Journal of Computational and Applied Mathematics*, **52**, 1994, 337-51.

46. ———, Generic bifurcation of periodic points, *Trans. Amer. Math. Soc.*, 149, 95-107.
47. ———, Periodic orbits near infinity in the restricted N-body problem, *Celest. Mech.* **23**, 1981, 69-81.
48. ———, Periodic solutions of the N-body problem, *J.Diff. Eqs.* **39(1)**, 1981, 2-38.
49. ———, Scaling Hamiltonian systems, *SIAM J. of Math. Anal.* **15(5)**, 1984, 877-89.
50. ———, Symmetries and integrals in mechanics, *Dynamical Systems* (Ed. M. Peixoto), Academic Press, New York, 1973, 259-72.
51. K. R. Meyer and G. R. Hall, *An Introduction to Hamiltonian Dynamical Systems*, Springer-Verlag, New York, 1991.
52. K. R. Meyer and D. S. Schmidt, Hill's lunar equations and the three-body problem, *J. Diff. Eqs.* **44(2)**, 1982, 263-72.
53. ———, Periodic orbits near L_4 for mass ratios near the critical mass ratio of Routh, *Celest. Mech.* **4**, 1971, 99-109.
54. ———, from the restricted to the full three-body problem, preprint.
55. J. K. Moser, Regularization of Kepler's problem and the averaging method on a manifolds, *Comm. Pure Appl. Math.*, 23, 1970, 609-635.
56. F. R. Moulton, A class of periodic orbits of superior planets, *Trans. Amer. Math. Soc.* **13**, 1912, 96-108.
57. ———, A class of periodic solutions of the problem of three bodies with applications to lunar theory, *Trans. Amer. Math. Soc.* **7**, 1906, 537-577.
58. ———, *Periodic Orbits*, Carnegie Inst. of Washington, Washington, DC, 1920
59. ———, The straight line solutions of the problem of N bodies, *Ann. of Math.* **2(12)**, 1910, 1-17.
60. V. Nechvile, Sur une forme nouvelle des équations différentielles du problème restreint élliptique, *Compt. Rend.* **182**, 1926, 310-11.
61. E. Noether, Invariante Variationsprobleme, *Nachr. König. Gesell. Wissen. Göttingen, Math. Phys. Kl.*, 1918, 235-257. (English Trans. *Transport Theory and Stat. Phy.* , 1971, 186-207.
62. F. Pacella, Central configurations for the N-body problem via equivariant Morse theory, *Arch. Rat. Mech.* **97**, 1987,59-74.
63. J. Palmore, Measure of degenerate relative equilibria, I, *Ann. of Math.* **104**, 1976,421-429.
64. ———, Minimally classifying relative equilibria, *Letters in Math. Phys.* **I**, 1977, 395-399.
65. O. Perron, Neue periodische Lösungen des ebenen Drei und Mehrkörperproblem, *Math. Z.* **42**, 1937, 593-624.
66. H. Poincaré, *Les méthodes nouvelles de la mécanique céleste*, Gauthier-Villars, Paris, 1892.
67. H. Polard, *Mathematical Introduction to Celestial Mechanics*, Prentice-Hall, Englewood, NJ, 1966.
68. C. Pugh and C. Robinson, The C^1 closing lemma, including Hamiltonians, *Erg. Th. Dyn. Sys.*, 3, 1977, 261-313.
69. P. H. Rabinowitz, The prescribed energy problem for periodic solutions of Hamiltonian systems, *Hamiltonian Dynamical Systems*, (Ed K. Meyer and D. Saari), Amer. Math. Soc., Providence, RI, 1988, 183-192.
70. D. Saari, On the role and properties of n body central configurations, *Celest. Mech.* **21**, 1980, 9-20.
71. D. Saari and J. Xia, Off to infinite in finite time, *Notices of Amer. Math. Soc.*, 42(5), 1995, 538-546.

72. D. A. Sánchez, *Ordinary Differential Equations and Stability Theory*, W. H. Freeman, New York, 1968.
73. W. Scheibner, Satz aus der Störungstheorie, *Crelle J. Reine Angew. Math.* **65**, 1866, 291-97.
74. D. S. Schmidt, Periodic solutions near a resonant equilibrium of a Hamiltonian system, *Celest. Mech.*, 9, 1974, 91-103.
75. J. Schubart, Librations in the elliptic restricted problem of three bodies, *Mathematische Methoden der Himmelsmechanik und Astronautik*, (Ed. E. Stiefel),1966, 21-31.
76. ______, Numerische Aufsuchung Periodischer Lösungen im Dreikörperproblem, *Astron. Nachr.* **283**, 1956, 17-22.
77. ______, Zur Regularisierung des Zweierstosses in Dreikörperproblem, *Astron. Nachr.* **282**, 1956, 262-264.
78. A. Sergysels-Lamy, Existence of periodic orbits of the second kind in the elliptic restricted problem of three bodies, *Celest. Mech.* **11(1)**, 1975, 43-51.
79. P. Shelus, A two-parameter survey of periodic orbits in the restricted problem of three bodies, *Celest. Mech.* **5(4)**, 1972, 487-489.
80. C. L. Siegel, Über eine periodische Loesung im Dreikoerperproblem, *Math. Nachr.* **4**, 1951, 28-35.
81. C. L. Siegel and J. K. Moser, *Lectures on Celestial Mechanics*, Springer-Verlag, New York, 1971.
82. S. Smale, Topology and mechanics, *Invent. Math.* **10** (1970), 305-331.
83. ______, Topology and Mechanics. II. The planar n-body problem, *Invent. Math.* **11**, 1970, 45-64.
84. J. Soler, Analytic continuation of circular Kepler motion to the general three-dimensional three-body problem, *Advanced Series in Nonlinear Dynamics* 8, (Ed. E. A. Lacomba and J. Llibre), World Scientific, 1996, 343-356.
85. F. Spirig and J. Waldvogel, The three-body problem with two small masses: a singular-perturbation approach to the problem of Saturn's coorbiting satellites, *Stability of the Solar System and its Minor Natural and Artificial Bodies* (Ed. V. Szebehely), Reidel, 1985.
86. V. Szebehely, *Theory of Orbits*, Academic Press, New York, 1967.
87. V. Szebehely and G. Giacaglia, On the elliptic restricted problem of three bodies, *Astron. J.* **69**, , 1964, 230-235.
88. J. Waldvogel, Collision singularities in gravational problems, *Recent Advances in Dynamical Astronomy* (Eds. Tapley and Szebehely), Reidel, 1973.
89. A. Weinstein, Symplectic manifolds and their Lagrangian submanifolds, *Advances in Math.* **6(3)**, 1971, 329-346.
90. ______, Normal modes for nonlinear Hamiltonian systems, *Invent. Math.*, 20, 1973, 47-57.
91. A. Wintner, *Analytical Foundations of Celestial Mechanics*, Princeton Univ. Press, Princeton, N.J., 1941.
92. E. T. Whittaker, *A Treatise on the Analytical Dynamics of Particles and Rigid Bodies*, Cambridge Univ. Press, London/New York, 1970.

Index

Lecture Notes in Mathematics

For information about Vols. 1–1525
please contact your bookseller or Springer-Verlag

Vol. 1526: J. Azéma, P. A. Meyer, M. Yor (Eds.), Séminaire de Probabilités XXVI. X, 633 pages. 1992.

Vol. 1527: M. I. Freidlin, J.-F. Le Gall, Ecole d'Eté de Probabilités de Saint-Flour XX – 1990. Editor: P. L. Hennequin. VIII, 244 pages. 1992.

Vol. 1528: G. Isac, Complementarity Problems. VI, 297 pages. 1992.

Vol. 1529: J. van Neerven, The Adjoint of a Semigroup of Linear Operators. X, 195 pages. 1992.

Vol. 1530: J. G. Heywood, K. Masuda, R. Rautmann, S. A. Solonnikov (Eds.), The Navier-Stokes Equations II – Theory and Numerical Methods. IX, 322 pages. 1992.

Vol. 1531: M. Stoer, Design of Survivable Networks. IV, 206 pages. 1992.

Vol. 1532: J. F. Colombeau, Multiplication of Distributions. X, 184 pages. 1992.

Vol. 1533: P. Jipsen, H. Rose, Varieties of Lattices. X, 162 pages. 1992.

Vol. 1534: C. Greither, Cyclic Galois Extensions of Commutative Rings. X, 145 pages. 1992.

Vol. 1535: A. B. Evans, Orthomorphism Graphs of Groups. VIII, 114 pages. 1992.

Vol. 1536: M. K. Kwong, A. Zettl, Norm Inequalities for Derivatives and Differences. VII, 150 pages. 1992.

Vol. 1537: P. Fitzpatrick, M. Martelli, J. Mawhin, R. Nussbaum, Topological Methods for Ordinary Differential Equations. Montecatini Terme, 1991. Editors: M. Furi, P. Zecca. VII, 218 pages. 1993.

Vol. 1538: P.-A. Meyer, Quantum Probability for Probabilists. X, 287 pages. 1993.

Vol. 1539: M. Coornaert, A. Papadopoulos, Symbolic Dynamics and Hyperbolic Groups. VIII, 138 pages. 1993.

Vol. 1540: H. Komatsu (Ed.), Functional Analysis and Related Topics, 1991. Proceedings. XXI, 413 pages. 1993.

Vol. 1541: D. A. Dawson, B. Maisonneuve, J. Spencer, Ecole d' Eté de Probabilités de Saint-Flour XXI - 1991. Editor: P. L. Hennequin. VIII, 356 pages. 1993.

Vol. 1542: J.Fröhlich, Th.Kerler, Quantum Groups, Quantum Categories and Quantum Field Theory. VII, 431 pages. 1993.

Vol. 1543: A. L. Dontchev, T. Zolezzi, Well-Posed Optimization Problems. XII, 421 pages. 1993.

Vol. 1544: M.Schürmann, White Noise on Bialgebras. VII, 146 pages. 1993.

Vol. 1545: J. Morgan, K. O'Grady, Differential Topology of Complex Surfaces. VIII, 224 pages. 1993.

Vol. 1546: V. V. Kalashnikov, V. M. Zolotarev (Eds.), Stability Problems for Stochastic Models. Proceedings, 1991. VIII, 229 pages. 1993.

Vol. 1547: P. Harmand, D. Werner, W. Werner, M-ideals in Banach Spaces and Banach Algebras. VIII, 387 pages. 1993.

Vol. 1548: T. Urabe, Dynkin Graphs and Quadrilateral Singularities. VI, 233 pages. 1993.

Vol. 1549: G. Vainikko, Multidimensional Weakly Singular Integral Equations. XI, 159 pages. 1993.

Vol. 1550: A. A. Gonchar, E. B. Saff (Eds.), Methods of Approximation Theory in Complex Analysis and Mathematical Physics IV, 222 pages, 1993.

Vol. 1551: L. Arkeryd, P. L. Lions, P.A. Markowich, S.R. S. Varadhan. Nonequilibrium Problems in Many-Particle Systems. Montecatini, 1992. Editors: C. Cercignani, M. Pulvirenti. VII, 158 pages 1993.

Vol. 1552: J. Hilgert, K.-H. Neeb, Lie Semigroups and their Applications. XII, 315 pages. 1993.

Vol. 1553: J.-L- Colliot-Thélène, J. Kato, P. Vojta. Arithmetic Algebraic Geometry. Trento, 1991. Editor: E. Ballico. VII, 223 pages. 1993.

Vol. 1554: A. K. Lenstra, H. W. Lenstra, Jr. (Eds.), The Development of the Number Field Sieve. VIII, 131 pages. 1993.

Vol. 1555: O. Liess, Conical Refraction and Higher Microlocalization. X, 389 pages. 1993.

Vol. 1556: S. B. Kuksin, Nearly Integrable Infinite-Dimensional Hamiltonian Systems. XXVII, 101 pages. 1993.

Vol. 1557: J. Azéma, P. A. Meyer, M. Yor (Eds.), Séminaire de Probabilités XXVII. VI, 327 pages. 1993.

Vol. 1558: T. J. Bridges, J. E. Furter, Singularity Theory and Equivariant Symplectic Maps. VI, 226 pages. 1993.

Vol. 1559: V. G. Sprindžuk, Classical Diophantine Equations. XII, 228 pages. 1993.

Vol. 1560: T. Bartsch, Topological Methods for Variational Problems with Symmetries. X, 152 pages. 1993.

Vol. 1561: I. S. Molchanov, Limit Theorems for Unions of Random Closed Sets. X, 157 pages. 1993.

Vol. 1562: G. Harder, Eisensteinkohomologie und die Konstruktion gemischter Motive. XX, 184 pages. 1993.

Vol. 1563: E. Fabes, M. Fukushima, L. Gross, C. Kenig, M. Röckner, D. W. Stroock, Dirichlet Forms. Varenna, 1992. Editors: G. Dell'Antonio, U. Mosco. VII, 245 pages. 1993.

Vol. 1564: J. Jorgenson, S. Lang, Basic Analysis of Regularized Series and Products. IX, 122 pages. 1993.

Vol. 1565: L. Boutet de Monvel, C. De Concini, C. Procesi, P. Schapira, M. Vergne. D-modules, Representation Theory, and Quantum Groups. Venezia, 1992. Editors: G. Zampieri, A. D'Agnolo. VII, 217 pages. 1993.

Vol. 1566: B. Edixhoven, J.-H. Evertse (Eds.), Diophantine Approximation and Abelian Varieties. XIII, 127 pages. 1993.

Vol. 1567: R. L. Dobrushin, S. Kusuoka, Statistical Mechanics and Fractals. VII, 98 pages. 1993.

Vol. 1568: F. Weisz, Martingale Hardy Spaces and their Application in Fourier Analysis. VIII, 217 pages. 1994.

Vol. 1569: V. Totik, Weighted Approximation with Varying Weight. VI, 117 pages. 1994.

Vol. 1570: R. deLaubenfels, Existence Families, Functional Calculi and Evolution Equations. XV, 234 pages. 1994.

Vol. 1571: S. Yu. Pilyugin, The Space of Dynamical Systems with the C^0-Topology. X, 188 pages. 1994.

Vol. 1572: L. Göttsche, Hilbert Schemes of Zero-Dimensional Subschemes of Smooth Varieties. IX, 196 pages. 1994.

Vol. 1573: V. P. Havin, N. K. Nikolski (Eds.), Linear and Complex Analysis – Problem Book 3 – Part I. XXII, 489 pages. 1994.

Vol. 1574: V. P. Havin, N. K. Nikolski (Eds.), Linear and Complex Analysis – Problem Book 3 – Part II. XXII, 507 pages. 1994.

Vol. 1575: M. Mitrea, Clifford Wavelets, Singular Integrals, and Hardy Spaces. XI, 116 pages. 1994.

Vol. 1576: K. Kitahara, Spaces of Approximating Functions with Haar-Like Conditions. X, 110 pages. 1994.

Vol. 1577: N. Obata, White Noise Calculus and Fock Space. X, 183 pages. 1994.

Vol. 1578: J. Bernstein, V. Lunts, Equivariant Sheaves and Functors. V, 139 pages. 1994.

Vol. 1579: N. Kazamaki, Continuous Exponential Martingales and *BMO*. VII, 91 pages. 1994.

Vol. 1580: M. Milman, Extrapolation and Optimal Decompositions with Applications to Analysis. XI, 161 pages. 1994.

Vol. 1581: D. Bakry, R. D. Gill, S. A. Molchanov, Lectures on Probability Theory. Editor: P. Bernard. VIII, 420 pages. 1994.

Vol. 1582: W. Balser, From Divergent Power Series to Analytic Functions. X, 108 pages. 1994.

Vol. 1583: J. Azéma, P. A. Meyer, M. Yor (Eds.), Séminaire de Probabilités XXVIII. VI, 334 pages. 1994.

Vol. 1584: M. Brokate, N. Kenmochi, I. Müller, J. F. Rodriguez, C. Verdi, Phase Transitions and Hysteresis. Montecatini Terme, 1993. Editor: A. Visintin. VII, 291 pages. 1994.

Vol. 1585: G. Frey (Ed.), On Artin's Conjecture for Odd 2-dimensional Representations. VIII, 148 pages. 1994.

Vol. 1586: R. Nillsen, Difference Spaces and Invariant Linear Forms. XII, 186 pages. 1994.

Vol. 1587: N. Xi, Representations of Affine Hecke Algebras. VIII, 137 pages. 1994.

Vol. 1588: C. Scheiderer, Real and Étale Cohomology. XXIV, 273 pages. 1994.

Vol. 1589: J. Bellissard, M. Degli Esposti, G. Forni, S. Graffi, S. Isola, J. N. Mather, Transition to Chaos in Classical and Quantum Mechanics. Montecatini Terme, 1991. Editor: 2S. Graffi. VII, 192 pages. 1994.

Vol. 1590: P. M. Soardi, Potential Theory on Infinite Networks. VIII, 187 pages. 1994.

Vol. 1591: M. Abate, G. Patrizio, Finsler Metrics – A Global Approach. IX, 180 pages. 1994.

Vol. 1592: K. W. Breitung, Asymptotic Approximations for Probability Integrals. IX, 146 pages. 1994.

Vol. 1593: J. Jorgenson & S. Lang, D. Goldfeld, Explicit Formulas for Regularized Products and Series. VIII, 154 pages. 1994.

Vol. 1594: M. Green, J. Murre, C. Voisin, Algebraic Cycles and Hodge Theory. Torino, 1993. Editors: A. Albano, F. Bardelli. VII, 275 pages. 1994.

Vol. 1595: R.D.M. Accola, Topics in the Theory of Riemann Surfaces. IX, 105 pages. 1994.

Vol. 1596: L. Heindorf, L. B. Shapiro, Nearly Projective Boolean Algebras. X, 202 pages. 1994.

Vol. 1597: B. Herzog, Kodaira-Spencer Maps in Local Algebra. XVII, 176 pages. 1994.

Vol. 1598: J. Berndt, F. Tricerri, L. Vanhecke, Generalized Heisenberg Groups and Damek-Ricci Harmonic Spaces. VIII, 125 pages. 1995.

Vol. 1599: K. Johannson, Topology and Combinatorics of 3-Manifolds. XVIII, 446 pages. 1995.

Vol. 1600: W. Narkiewicz, Polynomial Mappings. VII, 130 pages. 1995.

Vol. 1601: A. Pott, Finite Geometry and Character Theory. VII, 181 pages. 1995.

Vol. 1602: J. Winkelmann, The Classification of Three-dimensional Homogeneous Complex Manifolds. XI, 230 pages. 1995.

Vol. 1603: V. Ene, Real Functions – Current Topics. XIII, 310 pages. 1995.

Vol. 1604: A. Huber, Mixed Motives and their Realization in Derived Categories. XV, 207 pages. 1995.

Vol. 1605: L. B. Wahlbin, Superconvergence in Galerkin Finite Element Methods. XI, 166 pages. 1995.

Vol. 1606: P.-D. Liu, M. Qian, Smooth Ergodic Theory of Random Dynamical Systems. XI, 221 pages. 1995.

Vol. 1607: G. Schwarz, Hodge Decomposition – A Method for Solving Boundary Value Problems. VII, 155 pages. 1995.

Vol. 1608: P. Biane, R. Durrett, Lectures on Probability Theory. Editor: P. Bernard. VII, 210 pages. 1995.

Vol. 1609: L. Arnold, C. Jones, K. Mischaikow, G. Raugel, Dynamical Systems. Montecatini Terme, 1994. Editor: R. Johnson. VIII, 329 pages. 1995.

Vol. 1610: A. S. Üstünel, An Introduction to Analysis on Wiener Space. X, 95 pages. 1995.

Vol. 1611: N. Knarr, Translation Planes. VI, 112 pages. 1995.

Vol. 1612: W. Kühnel, Tight Polyhedral Submanifolds and Tight Triangulations. VII, 122 pages. 1995.

Vol. 1613: J. Azéma, M. Emery, P. A. Meyer, M. Yor (Eds.), Séminaire de Probabilités XXIX. VI, 326 pages. 1995.

Vol. 1614: A. Koshelev, Regularity Problem for Quasilinear Elliptic and Parabolic Systems. XXI, 255 pages. 1995.

Vol. 1615: D. B. Massey, Le Cycles and Hypersurface Singularities. XI, 131 pages. 1995.

Vol. 1616: I. Moerdijk, Classifying Spaces and Classifying Topoi. VII, 94 pages. 1995.

Vol. 1617: V. Yurinsky, Sums and Gaussian Vectors. XI, 305 pages. 1995.

Vol. 1618: G. Pisier, Similarity Problems and Completely Bounded Maps. VII, 156 pages. 1996.

Vol. 1619: E. Landvogt, A Compactification of the Bruhat-Tits Building. VII, 152 pages. 1996.

Vol. 1620: R. Donagi, B. Dubrovin, E. Frenkel, E. Previato, Integrable Systems and Quantum Groups. Montecatini Terme, 1993. Editors:M. Francaviglia, S. Greco. VIII, 488 pages. 1996.

Vol. 1621: H. Bass, M. V. Otero-Espinar, D. N. Rockmore, C. P. L. Tresser, Cyclic Renormalization and Auto-morphism Groups of Rooted Trees. XXI, 136 pages. 1996.

Vol. 1622: E. D. Farjoun, Cellular Spaces, Null Spaces and Homotopy Localization. XIV, 199 pages. 1996.

Vol. 1623: H.P. Yap, Total Colourings of Graphs. VIII, 131 pages. 1996.

Vol. 1624: V. Brınzanescu, Holomorphic Vector Bundles over Compact Complex Surfaces. X, 170 pages. 1996.

Vol.1625: S. Lang, Topics in Cohomology of Groups. VII, 226 pages. 1996.

Vol. 1626: J. Azéma, M. Emery, M. Yor (Eds.), Séminaire de Probabilités XXX. VIII, 382 pages. 1996.

Vol. 1627: C. Graham, Th. G. Kurtz, S. Méléard, Ph. E. Protter, M. Pulvirenti, D. Talay, Probabilistic Models for Nonlinear Partial Differential Equations. Montecatini Terme, 1995. Editors: D. Talay, L. Tubaro. X, 301 pages. 1996.

Vol. 1628: P.-H. Zieschang, An Algebraic Approach to Association Schemes. XII, 189 pages. 1996.

Vol. 1629: J. D. Moore, Lectures on Seiberg-Witten Invariants. VII, 105 pages. 1996.

Vol. 1630: D. Neuenschwander, Probabilities on the Heisenberg Group: Limit Theorems and Brownian Motion. VIII, 139 pages. 1996.

Vol. 1631: K. Nishioka, Mahler Functions and Transcendence.VIII, 185 pages.1996.

Vol. 1632: A. Kushkuley, Z. Balanov, Geometric Methods in Degree Theory for Equivariant Maps. VII, 136 pages. 1996.

Vol.1633: H. Aikawa, M. Essén, Potential Theory – Selected Topics. IX, 200 pages.1996.

Vol. 1634: J. Xu, Flat Covers of Modules. IX, 161 pages. 1996.

Vol. 1635: E. Hebey, Sobolev Spaces on Riemannian Manifolds. X, 116 pages. 1996.

Vol. 1636: M. A. Marshall, Spaces of Orderings and Abstract Real Spectra. VI, 190 pages. 1996.

Vol. 1637: B. Hunt, The Geometry of some special Arithmetic Quotients. XIII, 332 pages. 1996.

Vol. 1638: P. Vanhaecke, Integrable Systems in the realm of Algebraic Geometry. VIII, 218 pages. 1996.

Vol. 1639: K. Dekimpe, Almost-Bieberbach Groups: Affine and Polynomial Structures. X, 259 pages. 1996.

Vol. 1640: G. Boillat, C. M. Dafermos, P. D. Lax, T. P. Liu, Recent Mathematical Methods in Nonlinear Wave Propagation. Montecatini Terme, 1994. Editor: T. Ruggeri. VII, 142 pages. 1996.

Vol. 1641: P. Abramenko, Twin Buildings and Applications to S-Arithmetic Groups. IX, 123 pages. 1996.

Vol. 1642: M. Puschnigg, Asymptotic Cyclic Cohomology. XXII, 138 pages. 1996.

Vol. 1643: J. Richter-Gebert, Realization Spaces of Polytopes. XI, 187 pages. 1996.

Vol. 1644: A. Adler, S. Ramanan, Moduli of Abelian Varieties. VI, 196 pages. 1996.

Vol. 1645: H. W. Broer, G. B. Huitema, M. B. Sevryuk, Quasi-Periodic Motions in Families of Dynamical Systems. XI, 195 pages. 1996.

Vol. 1646: J.-P. Demailly, T. Peternell, G. Tian, A. N. Tyurin, Transcendental Methods in Algebraic Geometry. Cetraro, 1994. Editors: F. Catanese, C. Ciliberto. VII, 257 pages. 1996.

Vol. 1647: D. Dias, P. Le Barz, Configuration Spaces over Hilbert Schemes and Applications. VII. 143 pages. 1996.

Vol. 1648: R. Dobrushin, P. Groeneboom, M. Ledoux, Lectures on Probability Theory and Statistics. Editor: P. Bernard. VIII, 300 pages. 1996.

Vol. 1649: S. Kumar, G. Laumon, U. Stuhler, Vector Bundles on Curves – New Directions. Cetraro, 1995. Editor: M. S. Narasimhan. VII, 193 pages. 1997.

Vol. 1650: J. Wildeshaus, Realizations of Polylogarithms. XI, 343 pages. 1997.

Vol. 1651: M. Drmota, R. F. Tichy, Sequences, Discrepancies and Applications. XIII, 503 pages. 1997.

Vol. 1652: S. Todorcevic, Topics in Topology. VIII, 153 pages. 1997.

Vol. 1653: R. Benedetti, C. Petronio, Branched Standard Spines of 3-manifolds. VIII, 132 pages. 1997.

Vol. 1654: R. W. Ghrist, P. J. Holmes, M. C. Sullivan, Knots and Links in Three-Dimensional Flows. X, 208 pages. 1997.

Vol. 1655: J. Azéma, M. Emery, M. Yor (Eds.), Séminaire de Probabilités XXXI. VIII, 329 pages. 1997.

Vol. 1656: B. Biais, T. Björk, J. Cvitanic, N. El Karoui, E. Jouini, J. C. Rochet, Financial Mathematics. Bressanone, 1996. Editor: W. J. Runggaldier. VII, 316 pages. 1997.

Vol. 1657: H. Reimann, The semi-simple zeta function of quaternionic Shimura varieties. IX, 143 pages. 1997.

Vol. 1658: A. Pumarino, J. A. Rodriguez, Coexistence and Persistence of Strange Attractors. VIII, 195 pages. 1997.

Vol. 1659: V, Kozlov, V. Maz'ya, Theory of a Higher-Order Sturm-Liouville Equation. XI, 140 pages. 1997.

Vol. 1660: M. Bardi, M. G. Crandall, L. C. Evans, H. M. Soner, P. E. Souganidis, Viscosity Solutions and Applications. Montecatini Terme, 1995. Editors: I. Capuzzo Dolcetta, P. L. Lions. IX, 259 pages. 1997.

Vol. 1661: A. Tralle, J. Oprea, Symplectic Manifolds with no Kähler Structure. VIII, 207 pages. 1997.

Vol. 1662: J. W. Rutter, Spaces of Homotopy Self-Equivalences – A Survey. IX, 170 pages. 1997.

Vol. 1663: Y. E. Karpeshina: Perturbation Theory for the Schrödinger Operator with a Periodic Potential. VII, 352 pages. 1997.

Vol. 1664: M. Väth, Ideal Spaces. V, 146 pages. 1997.

Vol. 1665: E. Giné, G. R. Grimmett, L. Saloff-Coste, Lectures on Probability Theory and Statistics 1996. Editor: P. Bernard. X, 424 pages, 1997.

Vol. 1666: M. van der Put, M. F. Singer, Galois Theory of Difference Equations. VII, 179 pages. 1997.

Vol. 1667: J. M. F. Castillo, M. González, Three-space Problems in Banach Space Theory. XII, 267 pages. 1997.

Vol. 1668: D. B. Dix, Large-Time Behavior of Solutions of Linear Dispersive Equations. XIV, 203 pages. 1997.

Vol. 1669: U. Kaiser, Link Theory in Manifolds. XIV, 167 pages. 1997.

Vol. 1670: J. W. Neuberger, Sobolev Gradients and Differential Equations. VIII, 150 pages. 1997.

Vol. 1671: S. Bouc, Green Functors and *G*-sets. VII, 342 pages. 1997.

Vol. 1672: S. Mandal, Projective Modules and Complete Intersections. VIII, 114 pages. 1997.

Vol. 1673: F. D. Grosshans, Algebraic Homogeneous Spaces and Invariant Theory. VI, 148 pages. 1997.

Vol. 1674: G. Klaas, C. R. Leedham-Green, W. Plesken, Linear Pro-*p*-Groups of Finite Width. VIII, 115 pages. 1997.

Vol. 1675: J. E. Yukich, Probability Theory of Classical Euclidean Optimization Problems. X, 152 pages. 1998.

Vol. 1676: P. Cembranos, J. Mendoza, Banach Spaces of Vector-Valued Functions. VIII, 118 pages. 1997.

Vol. 1677: N. Proskurin, Cubic Metaplectic Forms and Theta Functions. VIII, 196 pages. 1998.

Vol. 1678: O. Krupková, The Geometry of Ordinary Variational Equations. X, 251 pages. 1997.

Vol. 1679: K.-G. Grosse-Erdmann, The Blocking Technique. Weighted Mean Operators and Hardy's Inequality. IX, 114 pages. 1998.

Vol. 1680: K.-Z. Li, F. Oort, Moduli of Supersingular Abelian Varieties. V, 116 pages. 1998.

Vol. 1681: G. J. Wirsching, The Dynamical System Generated by the 3n+1 Function. VII, 158 pages. 1998.

Vol. 1682: H.-D. Alber, Materials with Memory. X, 166 pages. 1998.

Vol. 1683: A. Pomp, The Boundary-Domain Integral Method for Elliptic Systems. XVI, 163 pages. 1998.

Vol. 1684: C. A. Berenstein, P. F. Ebenfelt, S. G. Gindikin, S. Helgason, A. E. Tumanov, Integral Geometry, Radon Transforms and Complex Analysis. Firenze, 1996. Editors: E. Casadio Tarabusi, M. A. Picardello, G. Zampieri. VII, 160 pages. 1998.

Vol. 1685: S. König, A. Zimmermann, Derived Equivalences for Group Rings. X, 146 pages. 1998.

Vol. 1686: J. Azéma, M. Émery, M. Ledoux, M. Yor (Eds.), Séminaire de Probabilités XXXII. VI, 440 pages. 1998.

Vol. 1687: F. Bornemann, Homogenization in Time of Singularly Perturbed Mechanical Systems. XII, 156 pages. 1998.

Vol. 1688: S. Assing, W. Schmidt, Continuous Strong Markov Processes in Dimension One. XII, 137 page. 1998.

Vol. 1689: W. Fulton, P. Pragacz, Schubert Varieties and Degeneracy Loci. XI, 148 pages. 1998.

Vol. 1690: M. T. Barlow, D. Nualart, Lectures on Probability Theory and Statistics. Editor: P. Bernard. VIII, 237 pages. 1998.

Vol. 1691: R. Bezrukavnikov, M. Finkelberg, V. Schechtman, Factorizable Sheaves and Quantum Groups. X, 282 pages. 1998.

Vol. 1692: T. M. W. Eyre, Quantum Stochastic Calculus and Representations of Lie Superalgebras. IX, 138 pages. 1998.

Vol. 1694: A. Braides, Approximation of Free-Discontinuity Problems. XI, 149 pages. 1998.

Vol. 1695: D. J. Hartfiel, Markov Set-Chains. VIII, 131 pages. 1998.

Vol. 1696: E. Bouscaren (Ed.): Model Theory and Algebraic Geometry. XV, 211 pages. 1998.

Vol. 1697: B. Cockburn, C. Johnson, C.-W. Shu, E. Tadmor, Advanced Numerical Approximation of Nonlinear Hyperbolic Equations. Cetraro, Italy, 1997. Editor: A. Quarteroni. VII, 390 pages. 1998.

Vol. 1698: M. Bhattacharjee, D. Macpherson, R. G. Möller, P. Neumann, Notes on Infinite Permutation Groups. XI, 202 pages. 1998.

Vol. 1699: A. Inoue,Tomita-Takesaki Theory in Algebras of Unbounded Operators. VIII, 241 pages. 1998.

Vol. 1700: W. A. Woyczyński, Burgers-KPZ Turbulence.XI, 318 pages. 1998.

Vol. 1701: Ti-Jun Xiao, J. Liang, The Cauchy Problem of Higher Order Abstract Differential Equations. XII, 302 pages. 1998.

Vol. 1702: J. Ma, J. Yong, Forward-Backward Stochastic Differential Equations and Their Applications. XIII, 270 pages. 1999.

Vol. 1703: R. M. Dudley, R. Norvaiša, Differentiability of Six Operators on Nonsmooth Functions and p-Variation. VIII, 272 pages. 1999.

Vol. 1704: H. Tamanoi, Elliptic Genera and Vertex Operator Super-Algebras. VI, 390 pages. 1999.

Vol. 1705: I. Nikolaev, E. Zhuzhoma, Flows in 2-dimensional Manifolds. XIX, 294 pages. 1999.

Vol. 1706: S. Yu. Pilyugin, Shadowing in Dynamical Systems. XVII, 271 pages. 1999.

Vol. 1707: R. Pytlak, Numerical Methods for Optical Control Problems with State Constraints. XV, 215 pages. 1999.

Vol. 1708: K. Zuo, Representations of Fundamental Groups of Algebraic Varieties. VII, 139 pages. 1999.

Vol. 1709: J. Azéma, M. Émery, M. Ledoux, M. Yor (Eds). Séminaire de Probabilités XXXIII. VIII, 418 pages. 1999.

Vol. 1710: M. Koecher, The Minnesota Notes on Jordan Algebras and Their Applications. IX, 173 pages. 1999.

Vol. 1711: W. Ricker, Operator Algebras Generated by Commuting Projections: A Vector Measure Approach. XVII, 159 pages. 1999.

Vol. 1712: N. Schwartz, J. J. Madden, Semi-algebraic Function Rings and Reflectors of Partially Ordered Rings. XI, 279 pages. 1999.

Vol. 1713: F. Bethuel, G. Huiksen, S. Müller, K. Steffen, Calculus of Variations and Geometric Evolution Problems. Cetraro, 1996. Editors: S. Hildebrandt, M. Struwe. VII, 293 pages. 1999.

Vol. 1714: O. Diekmann, R. Durrett, K. P. Hadeler, P. K. Maini, H. L. Smith, Mathematics Inspired by Biology. Martina Franca, 1997. Editors: V. Capasso, O. Diekmann. VII, 268 pages. 1999.

Vol. 1715: N. V. Krylov, M. Röckner, J. Zabczyk, Stochastic PDE's and Kolmogorov Equations in Infinite Dimensions. Cetraro, 1998. Editor: G. Da Prato. VIII, 239 pages. 1999.

Vol. 1716: J. Coates, R. Greenberg, K. A. Ribet, K. Rubin, Arithmetic Theory of Elliptic Curves. Cetraro, 1997. Editor: C. Viola. VIII, 260 pages. 1999.

Vol. 1717: J. Bertoin, F. Martinelli, Y. Peres, Lectures on Probability Theory and Statistics. Saint-Flour, 1997. Editor: P. Bernard. IX, 291 pages. 1999.

Vol. 1719: K. R. Meyer, Periodic Solutions of the N-Body Problem. IX, 144 pages. 1999.

4. Lecture Notes are printed by photo-offset from the master-copy delivered in camera-ready form by the authors. Springer-Verlag provides technical instructions for the preparation of manuscripts. Macro packages in TEX, LATEX2e, LATEX2.09 are available from Springer's web-pages at

http://www.springer.de/math/authors/b-tex.html.

Careful preparation of the manuscripts will help keep production time short and ensure satisfactory appearance of the finished book.

The actual production of a Lecture Notes volume takes approximately 12 weeks.

5. Authors receive a total of 50 free copies of their volume, but no royalties. They are entitled to a discount of 33.3 % on the price of Springer books purchase for their personal use, if ordering directly from Springer-Verlag.

Commitment to publish is made by letter of intent rather than by signing a formal contract. Springer-Verlag secures the copyright for each volume. Authors are free to reuse material contained in their LNM volumes in later publications: A brief written (or e-mail) request for formal permission is sufficient.

Addresses:

Professor F. Takens, Mathematisch Instituut,
Rijksuniversiteit Groningen, Postbus 800,
9700 AV Groningen, The Netherlands
E-mail: F.Takens@math.rug.nl

Professor B. Teissier, DMI, École Normale Supérieure
45, rue d'Ulm,
F-7500 Paris, France
E-mail: Teissier@ens.fr

Springer-Verlag, Mathematics Editorial, Tiergartenstr. 17,
D-69121 Heidelberg, Germany,
Tel.: *49 (6221) 487-701
Fax: *49 (6221) 487-355
E-mail: lnm@Springer.de

9 783540 666301